T0199337

DIVING

SEALS AND

MEDITATING

YOGIS

DIVING

SEALS AND

MEDITATING

YOGIS

Strategic Metabolic Retreats

Robert Elsner

The University of Chicago Press
Chicago and London

ROBERT ELSNER is professor emeritus at the University of Alaska Fairbanks. He studies the physiology of marine mammals and is the coauthor of *Diving and Asphyxia: A Comparative Study of Animals and Man*.

The University of Chicago Press, Chicago 60637
The University of Chicago Press, Ltd., London
© 2015 by The University of Chicago
All rights reserved. Published 2015.
Printed in the United States of America

24 23 22 21 20 19 18 17 16 15   1 2 3 4 5

ISBN-13: 978-0-226-24671-0 (cloth)
ISBN-13: 978-0-226-24704-5 (e-book)
DOI: 10.7208/chicago/9780226247045.001.0001

Library of Congress Cataloging-in-Publication Data

Elsner, Robert, author.
  Diving seals and meditating yogis : strategic metabolic retreats / Robert Elsner.
    pages   cm
  Includes bibliographical references and index.
  ISBN 978-0-226-24671-0 (cloth : alk. paper)
  ISBN 978-0-226-24704-5 (e-book)
  1. Seals (Animals)—Behavior. 2. Diving. 3. Metabolism—Regulation. I. Title.
  QL 737.P 6E 58 2015
  599.7915—dc23

2014044791

♾ This paper meets the requirements of ANSI/NISO Z 39.48-1992 (Permanence of Paper).

I gratefully dedicate this writing to my wife, Elizabeth, who provided support and helpful insights throughout its preparation.

CONTENTS

An appropriate route to new biological knowledge of animal physiology is recognized to be that of comparative studies of diverse species. But it is well to acknowledge the novelty of that rationale at its origin in the seventeenth century and its revolutionary departure from the rigid doctrine that had dominated thinking for centuries before that time. That new perspective on Nature was derived from quantitative experimental considerations, contrasting with the unchallenged acceptance of intellectual authority that had suppressed inquiry until this awakening. Daniel Boorstin's book *The Discoverers* (1983) relates the repressive theme by which the historical development of science had been inhibited by "the tyranny of Galen." Rigid interpretation of prevailing concepts as unchangeable truths severely limited objective research throughout that long period of submerged curiosity regarding how the natural world functions.

That restricted view of Nature was eventually breached by the enterprise of a few pioneers who recognized the conduct of experiments as an appropriate route to comprehension of the natural world. Prominent among this new generation was William Harvey (1578–1657), the originator of the conceptual basis of blood circulation, establishing animal physiology as a quantitative science. The accepted and unchallenged view of that time was that blood flow was restricted to back-and-forth motion within the heart and adjacent blood vessels. A break with previous centuries of biological mystery was heralded by Harvey's deductive experimental interpretation of the circulation of blood described in his landmark 1653 publication, *An Anatomical Disputation concerning the Movement of the Heart and Blood in Living Creatures* (original in Latin). It was Harvey's friend and colleague Robert Boyle, better known for the relationships governing gas volume, pressure, and temperature, who authored a pioneering discussion of human breath-hold diving (Boyle 1670). They shared an interest in exploring realities of the natural world that drove that intellectual renaissance.

Harvey's revolutionary clarification depended upon simple quantitative considerations relating to the amount of blood ejected by the heart at each beat and the consequent need to account for its circulation. His initiative reveals a resourceful and quantitative inquiry into living processes, a new approach and one that has guided inquiry since his time. The first chapter opens with the declaration: "When I first applied my mind to observation from the many dissections of living creatures as they came to hand, that by that means I might find out the use and benefits of the motion of the heart through actual inspection with my own eyes, and not from books and the writings of other men. . . ." Harvey's arguments depend on the innovative concept that blood is pumped by the heart "into the extremities back into the heart again and thus completes a movement as it were in a circle."

He was unaware of capillaries, since the invention of microscope technology for their detection was still a few years in the future, yet he hypothesized their existence as a rational explanation for the completion of the circulatory route. Harvey's original contributions to a new way of understanding life processes established the concept of studying animal and human reactions in the search for examples suitable for integrative study. To this end Harvey examined the heart and blood vessels of many living species, an approach that was to become the productive technique of physiological investigations. Thus opened a new era of research upon which our understanding and appreciation of the basis of animal adaptations depend.

Human responses to unusual environments are generally so obvious that we are well aware of their effects. Such reactions are especially noticeable when our comfort is disturbed and we need protection from exposure to threatening environments. Attention is demanded when we dive under water and are temporarily forced to stop breathing. Our curiosity may be aroused by contrasting our modest ability for breath-holding with the obvious and comfortable facility of seals and other marine mammals during their frequent and long immersions. These natives of aquatic habitats are well adapted for routine breath-holding immersions in pursuit of their regular lifestyles. Study of diving seals reveals the mechanisms upon which their underwater endurance depends and makes clearer the comparative reactions of other species, ourselves included.

Learning how animals adjust to different environments is accomplished by both field observation and laboratory study. Successful endurance of the marine mammal's long submersion depends upon its ability to temporarily depress the reactions essential for life. Similar profound metabolic inhibition is unusual in humans. Meditation as practiced by those engaged in its more dedicated expression provides a rare example of human metabolic suppression that can be made subject to voluntary control. The subjects presented in these pages concern comparisons of animal and human responses to these temporary metabolic retreats.

Discussions with colleagues and students have helped to clarify what I wish to convey regarding the physiological reactions being considered here. I am grateful for their help and for numerous fruitful exchanges with them. The working scientist is frequently reminded of the rich legacy that has inspired new inquiries and of the contributions of many investigators to the subjects of his interest. I owe a special debt of gratitude to the wisdom and inspiration of two mentoring colleagues, now deceased, with whom I have been privileged to work: Per Scholander, Scripps Institution of Oceanography, University of California, and Michael Daly, University of London.

I first encountered Scholander in the winter of 1942. He had recently arrived in the United States from Norway to work with Laurence Irving at Swarthmore College on physiological studies of marine mammals. Their work was interrupted by the imminence of World War II. They volunteered for military service and were assigned the task of developing equipment and techniques for protection of airmen forced into emergency survival conditions in harsh climates.

On this occasion they had chosen Mount Washington, New Hampshire, as their most readily accessible location for severe weather testing. They were accompanied by Irving's daughter, Susan, and George Llano, who was newly inducted into military service from graduate school. It was my good fortune to be working as a meteorological technician at the Mount Washington Observatory where we accommodated them.

Winter visitors were rare, and we were pleased by the opportunity to help in their projects. Pete and Larry, as we soon came to know Scholander and Irving, explained to us what environmental testing they wished to do. The hazard of carbon monoxide poisoning in emergency situations was of incidental interest, and for its testing Pete and I built a snow shelter within which we were to be exposed to the combustion products of small petroleum-burning stoves designed for military use. He had developed a portable kit for CO detection in a drop of blood, and that was to be the method for revealing its production by the stoves. Modest CO levels were indicated, but when a pan of snow was placed on the stove for melting to water, the resulting cooled combustion led abruptly to its profuse output and the prompt elevation of our blood levels. This observation was a clear lesson for the potential hazard of faulty ventilation while using such stoves.

I was fascinated by conversations with Pete in our snow shelter, especially his recounting of scientific adventures, and by participation with him and Larry in other projects. Careers changed abruptly during that wartime period; within a few months I was overseas driving an ambulance. The Irving-Scholander brand of science, the combination of field and laboratory investigations, had made a lasting impression on me, and my postwar return to education was much influenced by that encounter.

Scholander and Irving were distinguished for their wartime work in applying physiological skills to the solution of problems likely to be encountered during military operations. Pete was awarded a medal for his heroic rescue by parachute of the survivors of a crashed aircraft in Alaska. After the war they returned to research interests on the biological effects of environmental exposures and other subjects. Larry settled at the University of Alaska, Pete at the University of Oslo and at the University of California Scripps Institution of Oceanography.

I again met our Mount Washington visitors in connection with their studies of responses to cold exposure in native people in several parts of the world. In 1961 I joined Pete at Scripps and turned my interests to marine mammals and human divers. Pete and Susan had married since our original meeting. George Llano subsequently headed a branch of the National Science Foundation responsible for biological studies in Antarctica, and we met again in connection with my projects there. Larry and I met at the University of Alaska when I joined that faculty in 1973. I am forever grateful for the stimulation and pleasure of working with them.

My work with seals at Barrow, Alaska, was dependent upon the friendly cooperation and assistance of Iñupiat hunters, especially Charles and Harry Brower. Collection of tissue samples was authorized in a permit issued by the Office of Protected Species, U.S. National Marine Fisheries Service. Much of the research was supported by funding from the U.S. National Science Foundation and the National Institutes of Health. Antarctic support was provided by the NSF and the U.S. Navy and Air Force.

I am grateful for substantial editorial guidance provided by India Cooper and to Christie Henry, Amy Krynak, and Joel Score for helpful exchanges, all at the University of Chicago Press. Thanks to Dixon Jones, Scott Lonergan, Jeffrey Simonson, and Mark Vallarino, University of Alaska, for assistance with preparation of illustrations and computer manipulations of the text; to Dael Goodman for early suggestions; to my daughter, Wendy, for drawings of human and animal subjects; to my son, Peter, for supporting discussions; and to Craig Heller, Mike Castellini, Brian Barnes, Julie Hagelin, and my wife, Elizabeth, for their helpful reviews of various parts of the text. I am, of course, responsible for whatever questionable remarks remain in its present condition.

1 STRATEGIC METABOLIC RETREATS

The diverse world of animal adaptations includes many unanticipated relationships. One might not expect to find a common theme relating the physiological reactions of seals, marine mammals of the world's oceans, and meditating practitioners of yoga. To cast this improbable net even further afield, we might include hibernating animals, infants during birth, near-drowning victims and clams at low tide. The common threads linking this unlikely mix of animals and situations are the reactions that they share for tolerating unfavorable environments by lowering their energetic requirements and withdrawing into states of depressed metabolism. Scrutiny of these diverse examples reveals some suggestive insights into the biology of survival and well-being. Animals in these withdrawn states are less dependent upon their customary levels of oxygen consumption, temporarily lessening their need for that life-sustaining resource, later resuming normal activity when conditions become more favorable.

Some physiological responses experienced in deep meditation, much less examined than those of diving seals, suggest metabolic resemblances of these human reactions to those tested in more detail and described in marine mammals. The reactions and adjustments of the breath-holding underwater seal have been subjected to extensive examination by numerous investigators for several decades. In contrast, the yogi's physiological responses during deep meditation have received little scrutiny, and information regarding them rests on only a few recorded examinations. Such results as exist, however, show metabolic depressions to the likely lowest levels healthfully experienced by humans.

Meditation is a practice derived from ancient Hindu and Buddhist traditions and similar customs of other sects. Its modern derivatives continue in customs both secular and religious. Yoga, derived from oriental history lost in antiquity, has become popular in Western societies for its emphasis on physical exercises and postures, generally described as a form relating to *hatha* yoga, and for the stress-relieving aspects of regular withdrawal into periods of meditation. These may approach the depths of that experienced by dedicated practitioners of *raja* yoga, deep meditative states, some lasting several hours. Reports of these conditions assume something of an exotic nature, and some of the more unusual claims of contemporary yoga adherents have been regarded with skepticism and doubt. These include assertions of physiological reactions that appear to be in conflict with our understanding of how the human body functions. I have wished to approach the subject by seeking demonstrable realities that exist in practice and relating them to what we know of comparative physiological reactions in humans and other animals.

The results from studies of experienced meditators have revealed examples of profound metabolic suppression, unusual in contrast with other conditions such as quiet rest or sleep in which metabolism is more modestly reduced. The number of experimental results is small, but the observed effects deserve attention by virtue of the unusual extent of the metabolic decline. Like other adaptations of ancient religions, yoga has been subject to modifications and interpretations in more recent years. Its contemporary derivatives as they are practiced by Western advocates have been somewhat modified from the original Asian roots to better accommodate the requirements of modern lifestyles.

Both diving seals and meditating humans engage in situations that similarly result in temporary alterations of their regular physiological reactions. Some aspects of these effects, especially those concerned with respiration and metabolism, upon which mammalian life depends, are central to the account described here.

Life's requirements

Maintenance of a physiological steady state requires integration of primary regulatory functions, neural and endocrine, for maintaining equilibrium from disturbances that might be imposed by altered environments or by temporary encounters with challenging conditions such as those requiring suspension of breathing. Breath-hold diving with its inevitable progressive asphyxia, often accompanied by cold exposure and swimming exercise, comprises an assault on the ordinary homeostatic condition of the animal. These encounters, for which seals and other marine mammals are well adapted, alter the animal's resting equilibrium, the condition against which its adjustments to change are contrasted. Seals have been the animals of choice for many studies; populations of those regularly studied are abundant, and their reactions to diving are readily demonstrated.

Life for air-breathing organisms, seals included, is dependent upon reliably sustained respiratory function by which oxygen is obtained and carbon dioxide is eliminated. The early historic origin of the photosynthetic revolution of plant life resulted in the production of oxygen and the initiation of oxidative metabolism. Gradual dependence upon oxygenation for sustaining life of organisms in nucleated cells began more than a billion years ago and more than a billion years after the appearance of the earliest life forms. Organisms evolved with cell nuclei containing mitochondria, the sites of oxidative reactions providing energy for more active life. This resource ultimately became the metabolic base upon which the subsequent evolution of multicellular organisms began. But it didn't come easily, because the high reactivity of oxygen is such that it threatened to result in what has been described as the Oxygen Holocaust by Margulis and Sagan ([1986] 1997). Instead, making this resource available in support of evolving life was a crowning achievement in the path toward life as it came to be known.

The appearance of oxygen in ancient atmospheres, the product of plant photosynthesis, led to the origin of oxygen-based metabolism, greatly improving the efficiency with which life's chemical processes take place. Its atmospheric concentration steadily increased to the presently stable level of 21 percent. Metabolic activities were in this process expanded beyond glycolysis to oxidative phosphorylation, upon which much-enhanced energetic resource the explosive evolution of more complex species depended. That condition heralded extensive biological innovations, among them more vigorous activity, the development of multicellular organisms, and sexual reproduction.

Establishment of circulatory transport was the basis for growth beyond that of small organisms dependent upon simple diffusion for direct exchange of essential substances. The subsequent emergence of mammalian life and its dependence on blood-borne substances supporting oxidative metabolism was a product of this biological revolution. The growth of our knowledge relating to life's origin and the development of its many forms is engagingly told in *The Fifth Miracle* by Paul Davies (1999).

Breath-holding seals

The extraordinary variations with which animal adaptations are expressed give evidence of the vast range of environments in which life can be accommodated. Appropriately, the inquiring investigator looks across the broad spectrum of animal species for those features that provide useful examples of natural selection. Among these are the species that have been endowed by evolutionary processes for successfully living in watery habitats—the aquatic mammals and birds. Their lifestyles contrast markedly with those of terrestrial animals like us and appear in some respects to challenge the assumptions upon which mammalian existence depends. Some seal species can suspend breathing for more than an hour while diving deep in near-freezing temperatures and darkness, thriving in an aquatic world by reactions far beyond our abilities. It is the realm of the marine mammals: seals, dolphins, and whales. Their obvious facility and comfort represent distinct examples of animals that are well prepared for routine suspensions of respiration, in striking contrast with our intolerance of more than brief breath-holding submergence.

Abundant populations of seals and other marine mammals arouse our curiosity because their lifestyles differ markedly from those of familiar land-based animals, ourselves included. Contrasting with the obvious human deficiencies in aquatic environments are healthy populations of species comprising warm-blooded, air-breathing mammals that thrive in the oceans and coasts of northern and southern subpolar seas. Diversity in size, behavior, life history, and degree of adaptation characterize these animals. They appear to be as comfortable when submerged as their terrestrial cousins are above water, and they give no indication of being deterred by what appear to us to be formidable obstacles to easy living. Their natural dives vary greatly in duration and physiological challenge—many are brief and undemanding, others are long and deep.

The harbor seal (*Phoca vitulina*) is an abundant species throughout coastal North American seas. Their primary adaptations for life in water include enhanced respiratory reserves in support of their diving habit and subcutaneous fat insulation against heat loss. Marine mammals dive for food and to escape from disturbance at the surface. Their swimming, diving, and breath-holding skills derive from a variety of special attributes, anatomical, physiological, and behavioral. Figuring prominently in their successful adaptations for aquatic habitats are their variations and augmentations of terrestrial mammals' related respiratory and circulatory reactions. These modifications of the terrestrial mammal's working processes have major impacts on conserving and extending their metabolic reserves.

Breath-hold diving initiates cardiovascular reactions, specifically slowed heart rate and selectively reduced circulation that favors the central nervous system and the heart musculature. These responses lead to metabolic suppression in nonvital organs and conservation of the seal's enriched respiratory resources to support long underwater immersion while delaying the need to resume breathing. By this

1.1. Harbor seal (*Phoca vitulina*), North Atlantic and North Pacific Oceans. Drawing by Wendy Elsner.

means, marine mammals adjust to long dives with reactions that result in conservation of enhanced reserves and continuation of essential metabolic activity of the brain and heart. It is an adjustment to an environmental threat not well represented in humans but essential for life in the aquatic mammal specialists. It contrasts with hibernation, a longer-term energy-conserving metabolic retreat involving withdrawal by some species into a state of much-reduced metabolism and lowered body temperature in response to seasonal climate change and diminished food supply.

Metabolism at rest

The basal metabolic rate defines the level at which an animal consumes oxygen during wakeful minimum activity. That condition refers to the ordinarily lowest sustainable rate of metabolism during quiet rest and well after lingering increases related to food digestion.

Basal conditions, clearly defined and obtainable in human studies, are more uncertain in experiments with animals; resting metabolism is often a more applicable term.

There are animal conditions in which low metabolic levels are routine. One of these is sleep, the obligatory resting period of modestly depressed metabolism that results in an irregular and variable reduction in oxygen consumption. It amounts in humans to an approximately and variable 10 percent decline from the minimum awake value. Hibernation is a more profound and regulated response essential for the well-being of certain mammalian species when confronted with reduced feeding resources and cooling environments temporarily too severe for survival. They withdraw into metabolic depressions awaiting the return of more favorable environmental conditions. Consideration of these diverse appropriate examples provides insights into the range of physiological reactions that accompany such declines in metabolic activity.

Hibernating mammals—ground squirrels, bears, and others— avoid the hazards of seasonal food scarcity and thermal extremes by withdrawing from activity and entering into deep torpor, sometimes approaching freezing levels of body temperature with much-reduced metabolism and respiration. Most are of small size and therefore vulnerable to an excessively high rate of heat loss during cold exposure. Bears, the largest hibernators, endure long winter sleep that is accompanied by less decline in body temperature and metabolic rate than that experienced by the smaller species. Bats retreat from their customary nocturnal activity to rest by day, during which time they depress metabolism and become moderately hypothermic. Hummingbirds, chickadees, and others may become lethargic for a few hours of each night, conserving energy by decreasing their high metabolic rates and lowering body temperatures.

Other examples of metabolic depression include clams and mussels exposed at low tide, marine invertebrates that need to be immersed to obtain adequate respiratory exchanges. But when they are exposed to air they "clam up" and cease activity while waiting for the next incoming tidal immersion. Lungfish and desert frogs prevent dehydration during dry seasons by burrowing deep underground, where they rest in torpor waiting for the next rainfall, which might be months away. Freshwater turtles remain buried in pond mud in a condition of metabolic limbo throughout winter months. Depression of metabolic activity and tolerance of accumulating metabolic products are primary protective mechanisms for survival. Conversion to dependence upon a nonrespiratory resource, anaerobic metabolism, becomes their survival mode (review: Nilsson and Lutz 2004). Much longer and more severe vital suppressions, some lasting for centuries, are endured by bacterial spores, leaving them apparently lifeless but capable of revival when conditions are appropriate.

Typical metabolic reductions associated with these and other examples are depressions of energetic activity determined in the nonexercising condition. The mammalian capability for reduction of resting metabolism, the primary subject of our consideration here, is readily apparent. Except for the extreme low value of the small hibernators, they include a range of roughly 25 to 70 percent of their resting metabolic rates. Of these, some are the subjects of these pages, including the meditating human, hibernating bear, newborn mammal, and diving seal. Their similar responses of lowered metabolism suggest that they may share some regulatory reactions for coping with that condition, despite differences of magnitude and techniques among their characteristic tolerances of respiratory reduction or suspension.

All mammals at some times during their lives are likely to encounter disruptions of regular breathing. For coping with that possibility, they are endowed with essential emergency defenses against asphyxia, and these reactions are especially well developed in the marine mammal diving specialists. Their adaptations for living in water take several demonstrable forms, including enhanced oxygen capacity in blood and muscle and restricted blood flow during dives, except for that required for maintaining the integrity of the central nervous system. The characteristics and reactions of their diving habit are subjects of physiological interest related to their tolerance of long breath-holding and to increased pressure encountered during deep dives. Many seal dives are relatively brief, but the seal's tolerance of long submergences and the associated alterations of respiratory and blood gas contents has made these animals useful subjects for study. Learning how they perform their routine diving activities is accomplished by techniques for recording their reactions during free dives and by controlled experimental testing.

Our concern for consideration in these pages is those dives characterized by the quietly resting, nonswimming, submerged seal, the condition in which its resting metabolism is not increased by muscular exercise. It is a condition that is relatively undocumented, because it takes place while the seal is underwater, remote from observation and free to engage in whatever conditions of exercise or rest it may wish.

The seal and the yogi

However widespread the recourse to metabolic withdrawal may be among life forms that are adapted for that condition, it appears to be only modestly expressed among the ordinary responses of our human species. Instead, we must depend upon our wits, the endowment of a brain capacity that is designed for maximum use in problem-solving and imagination, for the invention of solutions to environmental threats. Depending upon how well we respond, whether involuntarily or subject to our curiosity for pushing life's boundaries, that capacity is successful in providing at least minimum survival reactions. It is unequal to the challenge when pushed too far into environmental extremes that surpass our regular response capacities. Endurance for breath-hold diving is such an example of modest capability among humans and other terrestrial mammals, in contrast to the long underwater endurances of the several species of seals, mammalian aquatic specialists.

Numerous levels of metabolic depression lie between those represented in various mammalian species by the modest decline during sleep and the extreme of hibernation. Two examples of reactions that depend upon such regulated suppressions are the breath-holding underwater submergence of the seal, specifically noteworthy during its resting, nonexercising condition, and the human reaction in a state of deep meditation. The choice of the submerged seal and the meditator for examination is based on their depressions of metabolism that represent the lowest levels of metabolic activity ordinarily achieved by seals and humans. Some aspects of their reactions suggest that similar

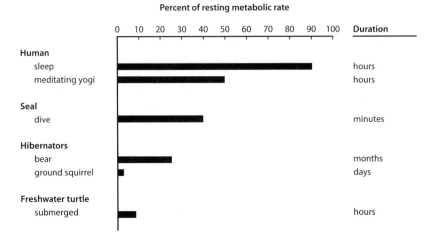

1.2.   Characteristic percent reduction and duration of resting metabolic rate in a variety of conditions and species. Responses of fetal and newborn mammals, preconditioned dog hearts, diving seals, and hibernating mammals are described in the text.

responses are evident in these disparate species and conditions of metabolic reduction, which they routinely experience as part of their regular lifestyles.

Meditation takes place in a fully conscious state in which metabolic activity is suppressed to values below those of sleep. Its physiological condition has been subject to little examination, and the evidence for its existence as a separate human state is based on modest and fragmentary evidence. That evidence, albeit meager, suggests that the metabolic depression achieved by dedicated and experienced practitioners of deep meditation may sometimes approach the depressed value that may occur in the quietly diving, nonexercising seal (figure 1.2). Based on the available modest direct evidence from meditating humans, they represent the nadir of metabolic reactions in both meditators and seals, declining to nearly one-half of the resting human metabolic rate and approaching anticipated similar and lower values in seals during quiet dives.

While quantitative characterizations of both meditating humans and nonexercising submerged seals exist in few recorded examples, it may nevertheless be useful to consider such evidence as exists for exploration of some indications regarding their responses in these circumstances.

Surviving asphyxia—lessons from seals

Diving in water puts obvious demands on the functioning and regulatory requirements of animals that depend upon air-breathing for their existence. However, for seals and other marine mammals, water is a productive and comfortable habitat, and sustained breath-holding is an obvious complement to that lifestyle. Many seal dives last no more than a few minutes, and their duration is reduced by the contradictory metabolic demands of swimming exercise. But when they choose to they can draw upon reserves that permit them to remain much longer underwater, especially when in a state of reduced activity and related lesser energetic demand.

Harbor seals (*Phoca vitulina*), common on the coasts of New England and the Pacific states, can dive for about twenty minutes; Antarctic Weddell seals (*Leptonychotes weddelli*) and northern and southern hemisphere elephant seals (*Mirounga angustirostris, Mirounga leonina*), for an hour or more. Brief submersions do not stress the animal's diving endurance. Seals' maximum abilities include long and deep excursions, and their lifestyles sometimes require them to approach the limits of their physiological resources.

Most species of dolphins and sea lions are relatively brief divers, generally no longer than a few minutes. But they appear to compensate for that breath-holding deficiency by being fast and agile swimmers, demonstrating a different lifestyle from that of the slower-swimming seals. Routine dives of whale species are highly variable, some long and deep, others comparatively brief and shallow. Sperm and bottlenose whales are long and deep divers, sometimes one hour duration and 1,000 meters depth. The relatively sluggish tropical sirenians (manatee and dugong) sustain regular immersions of several minutes.

From what we know of seal physiology, we can describe a predictable sequence of events occurring when seals submerge in water. Our knowledge regarding the physiological aspects of their diving is the result of extensive investigations of animals that are conveniently studied in their natural habitats and in captivity. When they dive and thereby cease breathing, often following pulmonary expiration, they are subjected to a steady depletion of oxygen present in the residual volume of their lungs, reserves bound to hemoglobin in circulating red blood cells, and that stored in muscle myoglobin. Both of these latter oxygen resources are well developed in seals; hemoglobin concentration is approximately twice the human value, and myoglobin is several times higher than in humans. Simultaneously, there is a corresponding cellular and blood increase in carbon dioxide and acidity throughout the dive, the by-products of metabolic activity. The change of concentrations—decreasing oxygen (hypoxia), increasing carbon dioxide (hypercapnia), and increasing hydrogen ions (acidosis), leading inexorably toward asphyxial conditions—is the unavoidable consequence of continued breath-holding. It is this triad combination of progressive asphyxia, not hypoxia alone, that best describes the condition of the breath-holding diver; its condition contrasts with the hypoxia, hypocapnia, and alkalosis encountered during high-altitude exposure.

The mechanisms that support the seal's tolerance of long breath-holding dives depend upon a well-regulated integration of physiological reactions associated with the distribution of respiratory gases via the circulation of blood. Responses to diving usually include a major reduction in heart rate and redistribution of blood flow. The circulation of blood is reduced, often by an abrupt beat-to-beat decline upon immersion, preferentially perfusing the brain at the expense of other body organs and tissues more tolerant of an interrupted blood supply. Heart rate falls, declining sometimes to one-tenth or less of the nondiving rate in reaction to the lessened demand for cardiac pumping. This response is the well-characterized diving bradycardia, the slowed pace of the heart's pumping action, noticeable during breath-holding immersion in virtually all vertebrates in which it has been looked for.

Reduction or cessation of blood flow, as occurs in the relatively hypoxia-tolerant intestines, kidneys, and other internal organs of diving seals, results in a precipitate lowering of metabolic activity in those organs. The advantages of depressed metabolism become especially apparent during long quiet dives in which competing metabolism of exercise is reduced. The energetic cost of swimming efforts regularly reduces dive durations to less than the quiet dive value, depending upon the intensity of the required muscular activity. The brain is particularly vulnerable to damage from oxygen deprivation, but the central nervous system of seals is notably more tolerant of asphyxial conditions than is the terrestrial mammal brain.

Toward the end of the seal's longest dives, its brain is perfused with blood containing much-reduced oxygen, reaching a minimum partial pressure of about 10 mm Hg (1.3 kPa), (Scholander 1940; Elsner, Shurley, Hammond, and Brooks 1970) well below the value, about 25 mm Hg (3.25 kPa), that can be expected to result in loss of consciousness in humans and other terrestrial mammals. In contrast, the nondiving, fully oxygenated seal's central nervous system functions with an oxygen pressure comparable to that of terrestrial mammals.

Human breath-holding is ordinarily limited to a comfortable duration of about one minute. Occasional reports of human diving exploits describe breath-holding duration of several minutes and depth recordings far in excess of the regularly demonstrated capacity. These are usually the result of purposeful, sometimes competitive, deep and rapid breathing in which lung and blood gas concentrations are altered by voluntarily increasing the rate and depth of breathing immediately before breath-holding The consequent alteration in concentrations of respiratory gases in blood circulating through the central nervous system can depress the subjective sense of air hunger that would normally arouse a demand for respiratory relief in the resumption of breathing. Notably, lung and blood levels of carbon dioxide and acidity that drive the sense of needing to breathe may in this condition be so diminished as to result in central nervous system hypoxia leading to loss of consciousness. This threat is especially pertinent during ascent from deep voluntary dives in which the respiratory gas pressures are more unfavorably altered by the accompanying reduction in ambient pressure, and loss of consciousness may be induced before reaching the surface. Prior oxygen breathing can similarly extend breath-holding endurance.

Mechanisms involved in the respiratory controls of the diver operate in a basically similar manner in seals and humans; their differing breath-holding abilities depend fundamentally upon quantitative rather than qualitative differences. The seal's obvious capacity for long immersions is governed by cardio-respiratory facilities that are common among mammals but are enhanced in the seal, an elaboration of the storage and regulatory mechanisms that govern respiration and circulation. They depend on reflex actions responding to activation of neural receptors located in the animal's face, pharynx, lungs, and receptors within the circulation that react to arterial pressure, blood gas tensions, and acidity. Our modest human diving capabilities are constrained by our reduced capacities that depend upon similar mechanisms but have much less reserve.

MARINE

MAMMAL

DIVERS

Knowing what marine mammals do and how they pursue their natural diving lifestyle occupies an obvious premier place in our perceptions of their activities in the aquatic environment. Equally important to our appreciation of their biology is an understanding of how their regulatory mechanisms operate to maintain effective responses and well-being. The two investigative purposes often require different methods and resources. Research approach to these differing goals will perforce employ multiple and wide-ranging techniques likely to shed light on whatever goals the investigator seeks to explore.

Much of our current knowledge has been derived from studies of several regionally abundant seal species, the marine mammal subjects considered here. Useful procedures vary from recording of free-ranging animal reactions to those undergoing rigidly controlled experimentation. Both techniques contribute to biological insights; both approaches are essential for discovering and appreciating their special aquatic adaptations. Laboratory experiments have usefully employed both experimental immersion trials and the training of animals to dive on signal. Several aspects of marine mammal diving performance have been subjected to study, but many questions remain for enterprising investigators. Seal reactions have been recorded in some extraordinarily long dives; equally impressive are exposures to increasing pressure with submergence, about one atmosphere for each ten meters of water depth. Seals of the family *Phocidae*, including harbor, ringed, gray, Weddell, hooded, and elephant seals, are among the marine mammal species whose diving habits and endurances have been examined. Diving capacities such as these animals have are subject to considerable modification by their age and by the nature of their underwater activity; swimming exercise can be expected to limit submerged endurance. Juvenile animals are generally brief and shallow divers (King 1983).

The diving reactions of several seal species among those abundant in the northern and southern hemispheres have been subjects of physiological studies. These include the following:

- Ringed seals (*Phoca hispida*) inhabit circumpolar regions of the Arctic Ocean.

- Harbor seals (*Phoca vitulina*) and spotted seals (*Phoca largha*) occur in the northern Pacific and Atlantic Oceans.

- Harp seals (*Pagophilus groenlandicus*) are found in the North Atlantic Ocean and the Arctic Ocean.

- Gray seals (*Halichoerus grypus*) exist in both sides of the northern Atlantic.

- Hooded seals (*Cystophora cristata*) are native to the sub-Arctic Atlantic Ocean.

- The Antarctic Weddell seal (*Leptonychotes weddelli*) exists in the circumpolar Antarctic.

- Elephant seals comprise two species: the northern (*Mirounga angustirostris*), of coastal California and Mexico, and the southern (*Mirounga leonina*), which frequents the seas near sub-Antarctic islands.

Seals range in adult body weight from about 45 kg (ringed seals) to 2,000 kg or more (male adult elephant seals). Adult female elephant

seals weigh substantially less than males, about 900 kg in both northern and southern species. The California sea lion (*Zalophus californianus*) has also been the subject of study. Adult males weigh about
250 kg and females 90 kg. They are relatively brief divers, generally
not more than a few minutes (Hurley and Costa 2001).

The diving exploits of some marine mammals have attracted particular notice. Maximum diving times of seals have been recorded to
range from about 15 minutes (ringed seals) to more than two hours
(northern elephant seals); depths range from 70 meters (ringed seals)
to 1,000 meters (hooded seals and elephant seals). Long-submerging
cetacean records include Cuvier's beaked whale's (*Ziphius cavirostris*)
submergence of 137 minutes and nearly 3,000 meters depth and the
sperm whale's (*Physeter macrocephalus*) submergence of similar duration and 2,250 meters depth.

The seal must depend upon its steadily declining respiratory resources throughout the dive. Cessation of breathing sets the course
toward progressive asphyxia, advancing inexorably unless reversed by
timely respiratory exchange. The overall effect limits the range, duration, and depth of underwater excursions, depending on the animal's
breath-holding endurance. The inevitable consequences of long dives
lead ultimately to the limiting of asphyxial tolerance, a condition recognized in the early studies of Scholander (1940) and by others since.
Elsner and Gooden (1983), Kooyman (1989), and Ramirez, Folkow,
and Blix (2007) review the systemic mechanisms for tolerating these
conditions by integrated metabolic responses of cellular and organ
systems.

The regular function of the circulation is the maintenance of
blood supply for tissues according to their metabolic needs, and organs vary widely in the immediacy of that required sustenance. An
obvious example is the occlusion of limb perfusion for many minutes
by a tourniquet, termination of blood flow resulting in a reduction
of metabolic activity in the muscle, skin, and other tissues of the limb.
Similar effects result from the abrupt decline of circulation as the
seal's dive begins. Oxygen conservation is achieved by limiting blood
flow to only the vital organs that are less tolerant of its temporary
decline. The competing demand for muscle blood supply when it
might be required in support of exercise will frequently set the limit

of dive duration, more exercise resulting in shorter dives. Metabolic suppression occurs in all except those organs in which perfusion is maintained, notably the brain and exercising muscle. Coronary blood flow, supporting the needs of heart muscle, may be reduced in accordance with the lessened need for circulation to tissues and organs whose supply is in less demand. Heart rate declines, responding to the decreased need for blood circulation.

The inexorable reduction of blood oxygen, progression of elevated carbon dioxide, and increased acidity of the blood and tissues dictate dive duration. This steady advance of hypoxia, hypercapnia, and acidosis signals the end point beyond which the demand for resumption of respiration becomes critically urgent, and the seal must surface for renewed respiration. Depressed metabolic demands of dives that involve little exercising activity, as in swimming and pursuit of prey, thus allow for longer submergence.

The seal's capability for quiet long-duration breath-holding, without the competing requirement for muscle blood supply in support of exercise, is the subject of my discussion here. Our limited knowledge of what happens during the seal's quiet dives with minimum or no exercise indicates that that condition permits maximum expression for whatever recourse to metabolic depression may serve to lengthen breath-holding duration. Deep excursions of some seals, for example, are performed by drifting with adjustment of buoyancy by emptying the lungs, thus reducing swimming effort during descent and at depth (Thompson and Fedak 1993; Williams et al. 2000). The nonexercising condition in these quiet dives reduces the seal's activity to an "idling" state characterized by lowering metabolism to its minimum value as required for long submersion.

The discourse here is devoted to a brief consideration of the physiological impacts encountered during diving and of the animal's responses to them. More detailed attention is given to these reactions in chapters 4 and 5.

Pioneer investigations
Origins of contemporary ideas regarding contrasts between diving mammal adaptations and those of terrestrial animals lacking their facil-

ity for enduring submergence are derived most notably from research begun in the 1930s and 1940s by Laurence Irving and Per Scholander. The concept that became a central theme in diving studies, differential circulation favoring the central nervous system, was proposed by Irving (1939). It was further elaborated in their joint publications, the products of experimental studies on seals and other marine mammals.

Recognition that the seal's capability for long dives is dependent upon other adaptations than only that of increased oxygen reserves in blood hemoglobin and the related oxygen-binding molecule in muscle, myoglobin, was one of the results of their pioneer studies. Oxygen stores alone, despite their augmentation in seals, are insufficient to support the animal's activity at a level of unchanged metabolism throughout long dives. In such conditions the cardiovascular system assumes an essential supporting function by reducing its circulatory distribution throughout the animal's body. Oxygen is thus conserved by preferential circulation to the seal's most vital organs, brain and heart, and by limiting blood flow in other organs that can better tolerate being temporarily deprived. Among these are kidney, gut, skin, and nonexercising muscle. The fundamental significance of this redistribution of circulating blood was considered in pioneer studies of Scholander (1940) as described in the monograph detailing his early laboratory studies on captive seals. He also noted the decline of oxygen consumption, along with a depression of body temperature that sometimes accompanied these dives. His experiments entailed ingenious use of such instrumentation as was available and, notably, Scholander's inventive skill for devising techniques as required for quantitative descriptions of respiratory and circulatory responses of experimentally dived animals.

Irving and Scholander's collaborative research provided the foundation for present-day approaches to understanding the physiology of diving. Their seal research anticipated the subsequent and current attention to comparative studies of diving mammals. Results of their productive enterprise are recorded in their publications of that period (Scholander, Irving, and Grinnell 1942a, 1942b; Irving, Scholander, and Grinnell 1942). The animals of their studies were primarily seals, but their wide-ranging interests also included dolphins and manatees. Among their important insights into how animals react to submergence was an appreciation for contributions of the seal's rich blood-borne oxygen reserves to supporting breath-holding dives by extensive alterations of cardiovascular reactions. The primary role of the cardiovascular system in the reactions of the seal's long dives was well established by clear lines of evidence resulting from their research and later studies to follow.

These experimental results were subsequently confirmed and extended by instrumentation of blood flow distribution in seals and recording of related events in other animals and humans (Elsner et al. 1966; Van Citters et al. 1965). The conspicuous attributes of the divers' asphyxial defenses have been subject to additional testing, verification, and review in several publications (Butler 2004; Butler and Jones 1982, 1997; Elsner and Gooden 1983; Ponganis, Kooyman, and Ridgway 2003; Davis et al. 2004; Castellini and Castellini 2004; Ramirez, Folkow, and Blix 2007; Zapol, 1996; Ponganis, Meir, and Williams 2011; Williams and Worthy 2000).

Circulation and metabolism

The various influences governing the diving seal's respiratory and circulatory reactions resemble those that operate in terrestrial mammals but with some essential quantitative elaborations of that model. The beginning of the dive and the associated cessation of breathing lead to slowing of heart rate, often abruptly within a few beats, and redistribution of the reduced cardiac output but with maintained circulatory perfusion of cardiac muscle and brain at reduced flow rates. Along with the decreases in circulation of organs better able to tolerate decreased blood flow, the seal's heart rate declines, demonstrating the characteristic diving bradycardia. Long dives impose a restriction of the blood supply that would be required by exercising muscle, resulting in increased dependence on anaerobic metabolic reserves, which are abundant in seals.

The contrast of diving seal responses with elevated mammalian cardiovascular demands of exercise is notable. Circulatory response to exercise requires increased cardiac output, with supportive increased myocardial blood flow for the supply of working muscle. Thus blood flow to working muscle is increased in humans, while kidney, gastrointestinal, and liver circulation are reduced (review: Rowell 1974).

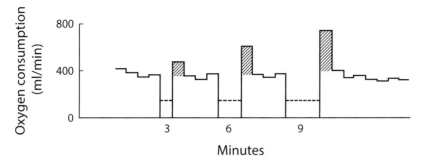

2.1.   Oxygen consumption before and after experimental dives of three, six, and nine minutes, gray seal (*Halichoerus grypus*). Crosshatched area, equivalent to area below dashed line, indicates extra oxygen consumed. Redrawn from Scholander (1940).

The resulting fractional redistribution of blood flow is qualitatively similar during exercise and diving; the restriction of circulation imposed during diving means that muscle work may depend more on anaerobic metabolic resources.

Metabolic activity decreases abruptly in the oxygen-deprived organs of the quietly resting diving animal. The reduced internal heat production and unavoidable heat loss to the surrounding water can lead to a decline of a few degrees in the seal's body temperature in lengthy dives and throughout a period of repeated dives (Scholander, Irving, and Grinnell 1942a, 1942b; Hill et al. 1987). This lowering of body temperature suppresses metabolic reactivity and, in combination with augmented oxygen stores in blood and muscle, permits longer breath-holding. The ensuing reduction in oxygen requirement of experimentally dived gray seals can amount to 60 to 70 percent less than that consumed in the nondiving condition (Scholander 1940; figure 2.1) and a further decline in the same species during nonexercising voluntary dives (Sparling and Fedak 2004).

The variation in these determinations is accounted for by differences in the test situations, experiments at rest contrasted with free-ranging conditions that may include exercise. California sea lions trained to dive were found to have 47 to 65 percent lower metabolic rates during quiet resting submersion than in the nondiving condition (Hurley and Costa 2001). A primary message from these observations is that metabolic depression is a regular feature of quiet, nonexercising dives. Metabolic requirements for the exercise of swimming impose a condition that competes for oxygen reserves, thus reducing the dive duration (Davis and Williams 2012).

Several circulatory adjustments combine to support the seal's dive endurance. The reduced coronary blood flow of the seal's heart is not maintained at a steady lower level; rather it is subject to frequent variations from ceasing entirely for brief intervals to high flow levels equivalent to those of the nondiving animal (Elsner et al. 1985). These circulatory variations within the myocardium appear likely to provide a means by which the seal heart's abundant glycogen content (Kerem, Hammond, and Elsner 1973) becomes available for continuing support of cardiac metabolism. The regular metabolic removal of potentially inhibiting acidic products assures the continued metabolism of the glycogen resource. Another exception to the blood flow restriction is the pregnant uterus, which is well supplied (Elsner, Hammond, and Parker 1970).

Diving experiments

Considerable variation exists among marine mammal species with respect to the extent and conditions of their dependence upon the characteristic cardiovascular responses during dives. These reactions of seals are also subject to modifications, varying with species and experimental conditions: free diving, trained dives, or forcibly induced immersions. Seal heart rates vary widely depending on the animal's condition, active, resting, diving, or combinations of these. Recordings from the free-swimming ringed seal (figure 2.2) and gray seal showed heart rate reductions to ten or fewer beats per minute during quiet, nonexercising unrestrained dives, increasing to one hundred or more during breathing intervals at the surface (Elsner et al. 1989; Thompson and Fedak 1993). Heart rates of free-diving Weddell seals are variable, sometimes thirty beats per minute in active dives lasting up to twenty minutes, slower in resting dives, and fifty to sixty while in the nondiving condition (Kooyman and Campbell 1973; Hill et al. 1987). Elephant seal heart rates range from ten beats per minute or less to about forty to forty-five per during free dives to about one hundred during nondiving rest (Andrews et al. 1997). Cardiac

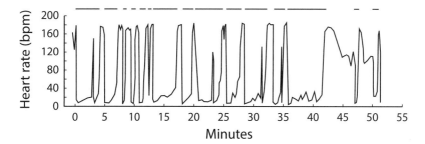

2.2. Heart rate variations of a captive ringed seal (*Phoca hispida*) free-diving in a pond near Fairbanks, Alaska. Submersion occurred during time indicated by horizontal line above. Redrawn from Elsner et al. 1988.

function responds acutely to the seal's dive situation, active or resting, sometimes declining to only a few beats per minute.

Many routine experimental dives have been performed by forcing the seal's immersion, whole body or head alone. The animal's response to this procedure results, not unexpectedly, in invoking intense protective reactions often greater than those that characterize its less threatening voluntary dives. The less traumatic procedures of recording from trained or free dives more nearly represent how the animal might respond in its natural diving conditions. Both approaches contribute to a full appreciation of what the seal does and what it can do. Desirable experimental procedures require close control sufficient for providing reliable insight regarding specific questions that the investigator may pose. Technological applications have succeeded in surmounting some of the problems and have made it possible to obtain physiological data from free-swimming animals. Useful interpretations have been derived from these studies, and comparisons with results from forced dives have helped to uncover the nature of diving reactions. Clearly, seals restrained and introduced directly into novel and threatening situations, such as forced immersions, can be expected to react with fear and resistance. That condition can be ameliorated by training in which the animal is introduced progressively and with minimum restraint to the experimental conditions until its composure is such that test immersions can be made with little obvious stress.

Operant conditioning techniques have been successfully employed in studies that depend upon training animals to perform discrete maneuvers, such as timed immersions. Animals that have been used in such studies have usually been in captivity for sufficient time to become well adjusted to the study team and procedures. This approach has been successfully exploited in controlled dives of California sea lions (Elsner, Franklin, and Van Citters 1964; Hurley and Costa 2001), harbor seals (Elsner 1965; Elsner et al. 1966; Jobsis, Ponganis, and Kooyman 2001), and bottlenose dolphins (*Tursiops truncatus*) (Elsner, Kenney, and Burgess 1966; Ridgway, Scronce, and Kanwisher 1969; Ridgway, Carter, and Clark 1975).

Neural regulations

Much of the present knowledge of the mechanisms that regulate the seal's diving reactions is derived from the comparative studies of respiration initiated by Daly and Angell-James (1975) in dogs and later extended to harbor seals in our collaborative studies. The seal reacts to immersion by a set of neural responses that have the effect of regulating the animal's endurance of long breath-holding. These include (1) inputs from facial skin receptors that respond to the presence of water by initiating respiratory and cardiovascular reflexes of slowed heart rate and much-reduced or arrested blood flow throughout the body, as well as maintained lower flow in brain and heart muscle; (2) neural signals from sensing elements (carotid bodies) responding to a steadily changing composition of gases dissolved in blood: decreased oxygen, increased carbon dioxide, and acidosis; and (3) resetting of arterial neural reactions for maintaining blood pressure (reviews: Daly 1984, 1997; Elsner and Daly 1988; Elsner 1999).

The fundamental elements of the seal's regulatory responses that initiate and modulate its responses during dives have been examined and compared with those of terrestrial mammals. They indicate that these initiating and sustaining control mechanisms that operate during dives are not unique to seals nor to the submerged condition. Rather, they represent considerably enhanced reactions of general mammalian neural reflex responses that govern the respiratory and circulatory systems. An examination of these governing mechanisms and some implications regarding their function are considered in another context in chapter 5.

The mammalian brain's intolerance of oxygen starvation plays a

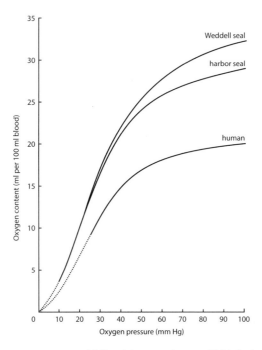

2.3.   Blood oxygen content, Weddell seal (*Leptonychotes weddelli*), harbor seal (*Phoca vitulina*), and human. Modified from Lenfant et al. 1969. Values indicated by dashed lines are inconsistent with consciousness.

primary role in setting the limit of diving endurance. The seal's brain, like that of other mammals, is sensitive to being deprived of oxygen-rich arterial blood and the accompanying removal of metabolic products via venous outflow. Brief shallow dives are sustained by oxygen withdrawn from its transit through the seal's lungs and blood. Blood oxygen capacity, the product of oxygen concentration bound to red cell hemoglobin and blood volume, is generally greater in seals than in terrestrial animals (figure 2.3). Blood volume, hemoglobin concentration, and red cell numbers are much increased in some species, thus enlarging the oxygen storage capacity of blood to several times greater than that of terrestrial mammals and playing an essential role in determining what the dive duration can be.

The oxygen and carbon dioxide in the respired air of mammalian lungs, conventionally expressed in terms of their partial pressures, PO_2 and PCO_2, are about 100 and 40 mm Hg (13 and 5.4 kPa) at a body temperature of 37°C and sea-level atmospheric pressure. The pH of mammalian arterial blood at normal body temperature is maintained at or near 7.4. This is a representative value for air-breathing mammals at rest, including marine species, progressing toward increasing blood acidity during breath-holding. Lengthy seal dives result in major alterations of these values, changes that at their extremes lie well outside of normal human blood pH values. In the seal's long dives arterial pH steadily declines, sometimes to 7.0 or less, a severely acidic condition for most mammals but one tolerated by seals.

Arterial oxygen and carbon dioxide levels during maximum sustainable Weddell and harbor seal dives may decline to partial pressures of 10 and 100 mm Hg (1.3 and 13 kPa), respectively (Elsner, Shurley, Hammond, and Brooks 1970). Hooded seals are similarly tolerant of extremely low brain oxygenation (Folkow et al. 2008). The seal's asphyxial tolerance is supported by its dense cerebral capillary network and the correspondingly shortened diffusion distances from blood to sensitive brain cells for exchanges of blood gases (Kerem and Elsner 1973a). In contrast, humans may lose consciousness when arterial oxygen partial pressure falls below about 25 to 30 mm Hg (3.3 to 4.0 kPa).

The seal's hypoxic tolerance was examined by Scholander (1940), who noted that gray and hooded seals survived extreme depletions of blood oxygen. The tolerance of the seal's critical organs, brain and heart, for exposure to the low oxygen levels encountered near the termination of long dives resembles the hibernating mammal's endurance of hypoxia. Chronic high-altitude exposure in mammals also results in development of cardiac resistance to hypoxia (Poupa et al. 1966; Ostadal, Ostadalova, and Dhalla 1999; Ostadal and Kolar 2007) and tolerance of cerebral structures to similar exposure. The low, and barely tolerable, human value of arterial oxygen pressure has been recorded in acclimatized mountain climbers at the summit of Everest: 28 mm Hg (3.7 kPa), in relatively alkaline blood resulting from loss of carbon dioxide by the hyperventilation that accompanies high altitude hypoxia (West et al. 1983).

Myoglobin, responsible for the oxygen store of skeletal muscle, exists in high concentration, ten or more times elevated in some seal species (Reed, Butler, and Fedak 1994; Kanatous et al. 1999; Polasek, Dickson, and Davis 2006; Ponganis et al. 2008; Burns et al. 2010) compared with its concentration in human muscle (Kagen and Christian 1966). An advantage of the high myoglobin concentration, in addition to oxygen storage, is its facilitated transport of oxygen and related support of aerobic metabolism (Hemmingsen 1963). Myoglobin is a likely contributor to the buffering of acid metabolites in skeletal muscle (Castellini and Somero 1981). Oxygen stores within blood and muscle of seals increase during maturity (Burns et al. 2007). Another respiratory protein, neuroglobin, has recently been identified (Burmester and Hankeln 2004). It exists primarily in the nervous sytem and appears to play a contributing role in the seal's tolerance of extreme brain hypoxia in long dives (Williams et al. 2008; Mitz et al. 2009).

Elephant seals, the largest and reportedly the longest and deepest seal divers, often dive repeatedly for long periods. Their diving habit sometimes allows only a few minutes for recovery at the surface (Hindell et al. 1992; Hindell, Slip, and Burton 1991). These sequentially repeated dives are typical of mature elephant seals during foraging at sea. The preferential circulation of blood to the central nervous system may give an advantage to large seals, because the brain characteristically represents a smaller fraction of body weight in related larger animals. The elephant seal's great size and consequent lower metabolic rate per kilogram of body weight can be expected to provide for a relatively increased oxygen reserve for the central nervous system and thus longer resistance to asphyxia.

The details of the elephant seal's successful and frequent long dives with only brief recovery at the surface are not well understood. Their brief surface intervals after long dives contrast notably with the extended times required by Weddell and other seals. The accompanying decline in temperature of a few degrees in some parts of the animal's body, noted in some seal species, can be expected to depress oxygen demand during repeated dives. This modification of body heat, originally noted in gray seals by Scholander (1940), has since been observed in other species (review: Ramirez, Folkow, and Blix 2007).

Conflicting metabolic demands

The seal's ability to sustain long dives is roughly related to body size. Both northern and southern adult elephant seals endure dives exceeding one hour, the longest of any seals. They tolerate longer dives than do smaller seals, a capability that may be related to the relatively small ratio of brain weight to body weight characteristic of large animals compared with smaller species (Crile and Quiring 1940; Von Bonin 1937). This relationship holds generally for species of the same family. The gray seal is noted for its frequent habitual quiet dives of up to twenty-five minutes, apparently a wait-and-capture feeding technique (Thompson and Fedak 1993). During such episodes its heart rate declines for occasional extended periods to fewer than four beats per minute, the lowest rate recorded in seals.

Breath-holding for sustained underwater endurance, as required for long dives, is clearly contrary to a need for sustained exercise (Castellini et al. 1985). When the submerged seal needs to swim vigorously in pursuit of prey, its demand for increased metabolic support of muscular effort and its underwater endurance are in direct conflict. Exercise requires skeletal muscle work supported by increased circulatory supply to active muscle and sustained by both aerobic and anaerobic metabolic resources.

These responses directly counter the requirement for cardiovascular suppression in lengthy dives and restrict the submerged seal's options for dive duration. The consequent conflict has been considered by Guppy et al. (1986) and Hotchachka and Guppy (1987), and they point out two cardiovascular responses derived from reported experimental results: the reduced heart rate and restriction of peripheral blood flow are graded depending upon the intensity of exercise. This issue has been further examined by Davis and Williams (2012) in a study in which the conflicting requirements of diving and muscular exercise of Weddell seals and bottlenose dolphins were tested. Cardiovascular support of exercise was found to vary inversely with the level of activity. The topic has been reviewed by Davis (2014), particularly in an examination of Weddell seal dives of varying exercise intensity as it relates to diving activity supported by oxygen-based (aerobic) metabolism. His detailed review of the topic demonstrates the need for recognizing that the responses of diving seals are subject to considerable modification depending upon the extent to which they may include reactions to swimming exercise coincident with the dive. The distinction is often hard to ascertain in experimental conditions in which the diving seal's activity, or lack thereof, may be difficult or impossible to observe or control.

The seal's capacity for maintaining aerobic exercise is relatively modest compared with the capability of active dogs and other endurance athletes. A requirement for exercise, as in pursuit of prey, imposes a challenge, the need for increasing oxygen consumption just while its available supply is steadily diminishing. The result is a conflicting disposition of the seal's metabolic resources and a corresponding abbreviation of dive duration (Davis et al. 2004). Metabolism in exercising seals is supported by the favorable distribution of mitochondria within the muscle cell (Watson et al. 2007) and by enhanced oxygen contents bound to muscle myoglobin and hemoglobin in circulating blood.

The seal's capacity for aerobic exercise is limited by its relatively low maximum oxygen consumption (Ashwell-Erickson and Elsner 1980; Sordahl, Mueller, and Elsner 1983; Elsner 1986), approximately five times its level at rest in harbor seals. The active human's value may be ten or more times resting, substantially more in athletes, higher still in dogs. But the seal can continue exercise well beyond its aerobic limit by recruiting its heightened capacity for anaerobic metabolism. Contribution from that source is substantially less in terrestrial animals, humans included.

The concept of altered circulation of blood in favor of nervous tissues and cells that have less resistance to the effects of ischemia has developed from several sources. It represents a fundamental characteristic of blood flow distribution in response to a variety of conditions that pose threats to cellular integrity. The likely first recorded instance of major circulatory alteration in a diving mammal is G. W. Steller's 1751 description of a hunting wound to the body of the now extinct Steller's sea cow (*Hydrodamalis gigas*) of the North Pacific Ocean: "As long as he kept his head under water the blood did not flow out, but as soon as he raised his head to breathe the blood leaped forth anew."

The overall effect of the cardiovascular responses to diving is the isolation of those portions of the seal's body that can better endure oxygen lack while adequate reserve supply is directed to the brain and heart. The net result is compartmental vasoconstriction and reduced cardiac pumping work. Skeletal muscle circulation is more labile, increasing blood flow when exercise requires it. This effective mechanism for conserving oxygen and reducing the contrary impact of increased metabolic products serves to maintain diving performance and duration. These reactions, well developed in the diving species, favor the oxygen-sensitive brain. They are not unique to those species; rather they are an enhancement of the general vertebrate circulatory design that provides metabolic protection for that vital organ.

The brain and other components of the mammalian central nervous system are particularly sensitive to interruption of regular circulatory sustenance. While the seal's brain is similarly the organ least tolerant of asphyxia, resistance to its adverse effects surpasses that of other comparably active and alert mammals. The inescapable question relates to the drastic reduction of circulation and concerns the possible disruption of function in affected regions, especially the central nervous system. Cellular elements of the seal's brain are prepared for this possibility by capacity for sustained function in conditions of long breath-holding, additionally showing unusually increased capability for anaerobic metabolism (Kerem and Elsner 1973a).

Physiological oxidations provide the required energetic base upon which life depends. But this process, however essential in support of life's complexities, carries with it the potential menace of the associated production of small amounts of highly toxic by-products. The acknowledged benefits of oxygen-based metabolism are accompanied by generation of small amounts of oxygen-derived free radicals and other intermediates of a toxic nature. They are produced especially in the circumstances of exposure to successive alterations of oxygenation and blood flow such as those associated with circulatory changes related to diving intervals. Detoxification depends on cellular modification of their potential effects (McCord 1985). Seals have been shown to have enhanced cellular protective mechanisms that counter these effects (Elsner et al. 1998; Johnson, Elsner, and Zenteno-Savín 2004; Vázquez-Medina, Zenteno-Savín, and Elsner 2006).

Body temperature and blood viscosity
The challenge of regulating body temperatures while immersed in near-freezing water imposes increased demands upon the marine

mammal's circulation of blood. Peripheral regions, flippers and fins, inevitably fall to temperatures close to those of the surrounding water, resulting in increased blood viscosity and consequent restrictions of blood flow. A further influence on circulation, the high concentration of red blood cells, (elevated hematocrit) notable in seal species, raises the question of a likely increased resistance to blood flow attributable to the related increase in viscosity (Wickham et al. 1990). The potential resulting effect is one of impaired circulatory perfusion and nutrient supply.

The comparative flow properties of blood from ringed, Weddell, and elephant seals have been examined (Meiselman, Castellini, and Elsner 1992). Of the three species studied, ringed seals are relatively brief divers, fifteen minutes being their usual longest duration, while Weddell and elephant seal maximum dive times are sometimes longer than one hour and two hours, respectively. Comparisons of blood flow characteristics among these species and with those of other mammals reveal some striking differences, but generalizations from the results are not readily apparent.

Weddell seals cope with the potential perfusion viscosity problem of high concentrations of red blood cells by sequestering them in the spleen and releasing them into the general circulation during dives (Hurford et al. 1996). Elephant seals, also very long divers, similarly resort to this technique. Such storage sequestration exists in those species in which the oxygen demands of long dives are supported by intermittent release of blood cells. This mechanism is not as apparent in species (ringed seals, for example) in which viscosity is lower by virtue of lesser red cell aggregation (Wickham et al. 1990). Relative spleen size in the contracted condition is considerably larger in Weddell seals than in ringed seals (0.9 percent vs. 0.2 percent of body weight). Natural selection has endowed these two species, Weddell and ringed seals, with differing mechanisms for dealing with oxygen conservation for long dives and oxygen delivery for sustained muscular effort. Temporary sequestration of red cells in Weddell seals and reduced red cell aggregation in ringed seals achieve similar results: reducing blood viscosity while not sacrificing oxygen storage, evolutionary solutions of a common problem by differing adaptive mechanisms.

Hypothermia can have a noticeable effect in reducing the seal's metabolic costs during diving, roughly one-half decline in metabolic rate per 10°C reduction in body temperature, thereby lengthening potential dive time. Body temperature of seals declines during dives, 2°C to 3°C in harbor seals (Scholander, Irving, and Grinnell 1942b; Hammel et al. 1977) and Weddell seals (Kooyman et al. 1980; Hill et al. 1987).

Experimental techniques

The pioneering studies initiated in the 1930s and 1940s by Irving and Scholander were directed primarily to experimental manipulation of restrained seals. This procedure allows the application of experiments for recording and manipulation of animal responses under controlled conditions. Restraint, however, influences the animals' reactions, yielding data relating to relatively extreme reactions, including maximum circulatory responses of bradycardia and restrictions of peripheral circulation. The procedure is thus more likely to respond to what the animal can do, rather than to what it routinely does, thus identifying its responses during limiting and extreme conditions. Differences in cardiovascular responses among seal species have been noted; their relative vulnerabilities to challenging dive and exercise circumstances are likely related to those differences.

Gerald Kooyman identified the variety and character of near-natural dives of Antarctic Weddell seals by the technique of instrument retrieval from unrestrained animals. The procedure depends on the benign attachment of time-depth recorders to the seal and the animal's release into an intentionally drilled hole in sea ice. Recovery of the instrumentation and recorded information is accomplished upon return of the swimming seal to the release location, assured in turn by its highly developed under-ice navigation skills and the planned location of that hole at a site distant from other potential breathing holes. This procedure and its derivatives have led to the collection of data relating duration and depth from thousands of dives (Kooyman 1989). The technique reveals much of the nature of free dives. It also assures a more natural experimental setting than that imposed by trained and forced dives.

Diving studies based on this approach have produced a wealth of data and appreciation of the physiological demands of diving in Weddell seals. These pioneer studies have resulted in important interpretations of the Weddell seal's lifestyle. Dives of this animal could be roughly characterized as brief—durations not exceeding fifteen to twenty minutes, the apparent limit of readily useable stored aerobic resources—and less frequent long dives that extend the seal's resources (Kooyman 1968, 1989). These dives are well short of the Weddell seal's maximum recorded durations, more than eighty minutes in this species (Castellini, Kooyman, and Ponganis 1992).

Among the results from studies undertaken with this technique are new understandings of the relative contributions of aerobic and anaerobic metabolic processes in support of the seal's diving routine (Kooyman et al. 1980; Kooyman 1989). The sequence of dependence upon these resources is such that much of the oxygen reserve in blood hemoglobin may be consumed before conversion to anaerobic metabolism. In Kooyman's (1968) words: "The advantage of an aerobic diving schedule is that no anaerobic metabolites, which require a long time to process, are accumulated, and there is little acid-base disruption. Instead, only the oxygen store needs to be replenished, a process that can be rapidly accomplished."

A later study (Kooyman et al. 1973) added further clarification, epitomized in this quote: "It has been noted that the average swimming metabolism of the Weddell seals was lower than resting metabolism. The authors were perplexed by this and suggested that perhaps the oxygen debt incurred on each dive was deferred in its repayment until a prolonged rest period or that other anaerobic metabolites were produced that were not strictly recoverable as an oxygen debt after the dive. It had been noted before in seals that the oxygen debt measured during recovery is less than the calculated oxygen deficit (Scholander 1940)."

Instrument-carrying seals

The technique of instrument recovery from seals returning to breathing holes in ice after dives was developed and exploited by Zapol and colleagues (reviews: Zapol 1996, 1987). They devised elaborate instrumentation for recording and blood sampling of Weddell seals at various stages of the dive.

These studies revealed much new information regarding the metabolic changes in blood gases, hemoglobin, and pH of arterial blood (Qvist et al. 1986). Notable among these results was the determination that circulating hemoglobin values increased substantially during long dives. The suspected source, later confirmed, was the contraction of the seal's spleen. The consequent likely increase of blood viscosity fit well the concept that the beneficial source of increased blood oxygen is thus made available during the dive by release from the spleen of sequestered red cells.

The publications by this group of investigators include several reports regarding Weddell seal responses recorded by computer-driven techniques for the study of cardiovascular and pulmonary reactions during unrestrained dives. Notable among their publications are evidence of seal lung collapse during deep dives (Falke et al. 1985) and regional blood flow during simulated dives (Liggins et al. 1980; Qvist et al. 1986; Zapol et al. 1979). A quotation from the Qvist et al. (1986) publication indicates the useful technology that was applied to the collection of physiological data from the freely diving seal, an innovative approach to the study of marine mammal physiology: "Our ability to sample aortic blood and to measure heart rate, depth, and temperature of the free-diving Weddell seal has allowed us to obtain new information about how this seal dives for such long periods. Our most important discovery was to uncover the Weddell seal's ability to consistently increase arterial hemoglobin and hematocrit during diving. This seal may inject oxygenated red blood cells to increase the central circulating arterial oxygen content without the disadvantage of circulating a large high-viscosity blood volume during rest periods."

The instrument recovery technique has been further elaborated and productive with recording from Weddell and elephant seals (Ponganis, Kooyman, and Ridgway 2003; Meir et al. 2009; Fedak 2009). Cardiovascular diving responses have been shown to differ among these species and those of other free-diving studies: gray seals (Fedak et al. 1988; Thompson and Fedak 1993) and ringed seals (Elsner et al. 1989). The gray seal's practice of extended periods of apparent quiet repose at depth is accompanied by extreme slowing of heart rate to four beats per minute and brief surface recoveries. The diving bradycardia of Weddell seals is less dramatic, and they require lengthy recovery before undertaking further long dives (Kooyman and Campbell 1973). It seems clear that diving reactions are substantially modified in different species and by the particular circumstances of their existence and lifestyles (Andrews et al. 1997).

MEDITATING

YOGIS

Comparisons of what seals and yogis do, diving and meditating, may appear to be an obscurely contrived exercise, an unlikely match of animal and human lifestyles. But examination of their characteristic responses during these conditions suggests that they share some similar and appropriate metabolic reactions. Our knowledge of how seals react to submersion, despite many remaining questions, is considerably more detailed and secure than the uncertainty of what we know about reactions of deeply meditating yogis and others who are devotees of this practice. Such fragmentary evidence as we have suggests that closer scrutiny is warranted. Resemblances to the metabolic suppression of the quietly diving seal suggest some similar responses.

Major metabolic declines are generally absent from the human adaptive repertoire. However, a closer look at the full scope of human responses reveals some exceptions to this condition. They are modest when compared with the deep metabolic depressions of some hibernators and diving mammals, but they are nevertheless readily identifiable and measurable. Experimental evidence suggests that certain human meditative practices result in physiological effects similar to the other mammalian responses to be considered here.

An extensive literature exists relating to the psychology of meditation (Naranjo and Ornstein 1971; Austin 1998, 2006; Rubia 2009; Lutz et al. 2008) but little that concerns the physiological responses that may accompany its practice. The scarcity of such information necessarily poses questions regarding what does exist as descriptions of responses during meditation. Regrettably, some yoga effects have been the subjects of exaggerated claims and counterclaims, indicating the need for cautious evaluation and consideration. The practice of meditation has been claimed to slow the rate of aging (Elissa et al. 2009). Serious concern for the study of various aspects of meditation indicates that the topic attracts increased attention, as indicated by recent publications of related conference proceedings (editors: Bushell, Olivo, and Theise 2009; Sequeira 2014). In recent years the more contemplative aspects of yoga have been adopted by devotees of Transcendental Meditation (TM) and related interests popular in the Western world. Their practice usually takes the form of daily meditation periods, episodes described as contemplative and refreshing departures from the more hectic and immediate concerns of contemporary life (Benson 1975).

Beyond its modest decline of about 10 percent during sleep, a further depression of the metabolic rate for whatever reason is noteworthy in humans due to its uncommon occurrence and the likely rare opportunity for its study. In contrast, submerged resting seals and sea lions are likely to routinely experience metabolic suppression to half or less of their nondiving value (Scholander 1940; Hurley and Costa 2001), and hibernating bears to one-quarter of their nonhibernating condition (Tøien et al. 2011). Comparable values of the much smaller, and accordingly more metabolically responsive, hibernating ground squirrel show declines to less than 5 percent of their nonhibernating condition (Barnes 1989). They experience multihour arousals from hibernation every few days, whereas bears may remain in torpor for weeks or months with only occasional postural movements. Black bears are closer to seals and humans in body weight and may be expected to more nearly approximate their resting metabolic rates. Examples of such metabolic decline in these and other animal responses argue for its having a role in adjusting to environmental impacts, and it can therefore be useful to examine the reactions of those species whose lifestyles involve its frequent expression. Recorded metabolic depressions suggest that it may also occur in those members of the human species who regularly practice another form of metabolic retreat, long and purposeful meditation.

Relief from the distractions of wandering thoughts and the relaxation of muscle tension are among the conceptual goals of meditation. Its practice is reported by devotees to result in decreased stress, improved powers of concentration, enhanced emotional stability, and a sense of well-being. These effects are sometimes described in subjective terms not readily interpretable in measured scientific discourse. As described in the *Encyclopaedia Britannica*: "The practice of meditation has occurred worldwide since ancient times in a variety of contexts. It may serve purely quietistic aims, as in the case of certain reclusive mystics; it may be viewed as spiritually or physically restorative and enriching to daily life, as in the case of numerous religious orders and the majority of secular practitioners; or it may serve as special, potent preparation for a particular, usually physical or otherwise strenuous activity, as in the case of the warrior before battle or the musician before performances."

Physiological responses

The physiological consequences of meditation are not inherently obvious; its practitioners describe it as a healthful expression of mental and emotional relaxation. This relaxation generally leads to reduced muscle tension, and this condition is reflected in slowed respiration and depressed metabolism. With practice, the intrusion of sensory stimulation is reduced, and a state of general calm ensues. If meditation takes place in a thermally neutral environment in which heat gain or loss is minimal, or in one that results in modest cooling, body heat storage and metabolic rate are gradually reduced. If sufficiently long lasting, this relaxed state may lead to a reduction of metabolic rate well below that which occurs during sleep. Results from the few reported pertinent physiological studies suggest that these thermal and metabolic responses regularly occur in well-practiced subjects during meditation. Western subjects customarily engage in relatively brief meditation periods, and it may therefore be difficult to obtain confirmation from them of the more profound responses characteristic of lengthier and deeper engagements. Nevertheless, the experimental results sometimes show metabolic reductions of an abrupt nature and of a greater magnitude than those recorded during other conscious human conditions.

Various aspects of the contemporary practices of meditation are examined in detail by Monaghan and Viereck (1999) and Broad (2012). Meditation as practiced in Western society is often limited to periods not longer than thirty minutes, terminated by increasing physical discomfort or boredom. Eyes may be open or closed, depending on the particular technique. Frequent episodes at convenient and regular times are recommended. Adherents customarily engage in daily meditative periods, often in the early morning. Some devotees participate in groups; others prefer to meditate alone. Quiet reflection is the goal. Some achieve it by focusing on repeated unspoken words or "mantras"; others prefer the visualization of mental images (Lutz et al. 2008). Brief sleeping moments may occasionally intervene, but the customary sitting or cross-legged postures limit or discourage them.

Some aspects of TM practice have been revealed by physiological examination, and these include prompt metabolic rate reduction that persists throughout the meditative episode. The onset of metabolic decline from the premeditative state is likely to be more abrupt than that which accompanies sleep, suggesting a direct regulatory influence that results from the training experience. Distinct and abrupt reductions in oxygen consumption to levels well below the approximately 10 percent decline of sleep have been demonstrated. These amount to regular declines of about 20 percent during meditation (Wallace 1970; Wallace, Benson, and Wilson 1971; Wallace and Benson 1972). A theme running through reports of its nature clearly indicate that meditation is a separate state distinguishable from sleep. However, the tendency for falling asleep is an ever-present distraction, especially among inexperienced meditators.

While much has been written about Eastern origins of meditation, research into physiological responses of its practitioners has received relatively little attention. Except for the modest reduction during sleep, humans are unlikely to experience major depressions of basal metabolic rate, the level at which the awake, resting adult functions in a thermally neutral environment. Indications of metabolic depression during transition to the meditating state are based on few experimental results. They suggest a change to a metabolic level well below that of sleep. The determinations that have been obtained from deeply meditating subjects, the lowest recorded in healthy humans, indicate suppressions declining nearly to one-half of the nonmeditating metabolism during quiet rest. They approximate values no more than twice those of hibernating bears (Tøien et al. 2011), surprising considering the bear's depressed metabolic state and lowered body temperature during that condition.

Historical perspective

Recourse to regular periods of quiet meditation can be traced into antiquity. Its expressions have a long history from origins in India, China, and Japan, and it has been associated, though not exclusively, with religious practice of diverse faiths. Some devotees adhere to a regular posture of sitting with erect back and supported by legs crossed in the "lotus" position or comfortable approximations of that condition (figure 3.1). Sitting upright with straight back and feet flat on the floor is an alternative mode for some practitioners. Whatever the posture, most devotees meditate in positions contrived to reduce the tendency for falling asleep.

The contemporary practice of Transcendental Meditation is a modern derivative of ancient prescriptions for recourse to conditions of physical rest and mental concentration. There have been few studies of its devotees concerned with physiological responses in their reactions to these episodes. Metabolic rate, as determined by oxygen consumption, decreased by 18 percent in nine experienced practitioners while meditating for thirty minutes (Wallace 1970). A subsequent study found a 16 percent reduction during twenty to thirty minutes of meditation (Wallace and Benson 1972). Similar metabolic

3.1.   Meditating yogi. Drawing by Wendy Elsner.

reductions are not usually encountered within this range in adult humans during ordinary sleep, an approximately 10 percent decline being the more usual condition.

Beary and Benson (1974) showed in another study that little special training was required to induce modest declines in oxygen consumption accompanying these brief meditative episodes. Volunteers were briefly instructed to relax while sitting quietly and, rather than specifically meditating, to concentrate on repeating, without vocalizing, a simple word (a "mantra") during expiration. This practice was accompanied by a decline of 13 percent in oxygen consumption. Differences from simple relaxed sitting were statistically significant. This and other observations have been reviewed by Davidson (1976).

Meditation and sleep are both conditions of reduced metabolic activity, but they differ in important respects. Meditative withdrawal bears some superficial resemblance to sleep, but that similarity is not indicative of comparable physiological conditions. Meditation is an awake conscious state in which the goal is a concentrated process of physical relaxation but with maintained mental alertness. Although brief dozing moments may be unavoidable, the more usual condition of the meditating subject is one of controlled relaxation incidental to the meditative state.

Clear descriptions are hard to come by. Perhaps a useful one is that of Huston Smith (1991). He contrasts the relative ease of achieving muscle relaxation is achieved with the difficulty of calming the more active mind: "At last the yogi is alone with his mind, but the battle is not yet won, for the mind's fiercest antagonist is itself. Closeted, it shows not the slightest inclination to settle down or obey. The yogi wants it to be still, to mirror reality in the way a quiet lake reflects the moon. Instead, its surface is choppy at best. Hinduism likens its restlessness to a crazed monkey."

Yogis, the practitioners of a discipline originating centuries ago in India, are better known in the West primarily for their practice of physical exercises that improve joint flexibility and muscle tone while lowering mental stress, collectively known as *hatha* yoga. Hindu and Buddhist tradition has been the philosophical source of meditating techniques as an expression of religious practice, daily meditation sometimes lasting several hours. It is regarded as an especially sublime state leading to the achievement of spiritual enlightenment, *raja* yoga. The rationale for these levels of practice depends upon their proclaimed ability for preventing intrusions upon the state of undisturbed contemplation. Thus, in simple terms, hatha yoga relates in part to conditioning of the physical body, while raja yoga is more concerned with the meditative state. Both aspects of yoga have attracted considerable attention in the Western world (Broad 2012), where they are practiced by devoted adherents. Koestler (1960) provides a Westerner's insightful impressions during encounters with its practice in India.

Dedicated Indian yogis engage in lengthy (one to several hours) daily meditations in the lotus posture or other comfortable positions. As an expression of the degree of control that they have attained over their bodies, some claim to be able through meditation to regulate autonomic functions such as heart rate and respiration, claims sometimes at odds with Western physiological discipline. In order to realize sensory deprivation leading to the deepest meditative states, some Indian yogis have themselves buried in isolation for hours or days in an

underground chamber. This practice is meant to symbolize the sub-
jugation of the body and freeing of the mind to engage in contemplation supported by withdrawal from interfering external influences.

Meditation tests

Research studies relating to yoga and similar philosophical practices
have been inspired by interest in Eastern philosophy and religion.
Physiological consideration of meditation practice has been limited
to a few experimental efforts, and the topic has attracted less attention
than it might deserve, considering the suggestive evidence for unusual
modification of ordinary function of the autonomic nervous system.
Experiments have been performed under widely varying conditions
and durations. Some have been devoted to a critical examination of
the reactions accompanying meditation such as that practiced by yogis. Suggestions of differences from normal human function implied
by these conditions involving few experimental subjects are deserving
of further investigative testing and experimental verification.

Considerable effort may be required to locate appropriate subjects
willing to participate in investigations of their practice and to accept
the strict protocols required. A few isolated physiological studies have
been undertaken with yogis who have a long history of devoted experience in meditative practice and who are willing to submit to
experimental study of their reactions. Variation exists among subjects,
depending upon their experience and the likely consequent limitation of responses. The full expression of physiological changes during
meditation is demonstrated in the few experimental results from trials
in which experienced Indian yogis and Western subjects have been
engaged. Depression of metabolism may be marked by an abrupt decline in oxygen consumption and maintenance of the lowered level
during the meditation period.

Meditation contrasts with the more irregular and modest metabolic depressions of sleep, including a lowering of about 1°C in deep
body temperature and response in average skin temperature, depending upon the thermal environmental (Kreider, Buskirk, and Bass 1958;
White, Weil, and Zwillich 1985). Thus the mean body temperature,
the weighted average of deep and surface temperatures, may some-

times remain little changed during undisturbed sleep. Some hormonal and neural responses are suggested by changes in blood levels of the pineal secretion melatonin and the neurotransmitter serotonin. They were found to be increased in practiced meditators (Solberg et al. 2004), possible contributors to a general elevation of mood and reduced anxiety.

A pioneer physiological study of meditation was undertaken by the Indian scientists Anand, Chhina, and Singh (1961). They tested the responses of a dedicated long-term practitioner of yoga during his confinement within a closed and sealed thermo-regulated box on two occasions for eight and ten hours. His oxygen consumption, calculated from the progressive change of gas composition within the box, declined during that time to a low of approximately 50 percent of his control value, the first demonstration of considerable reduction in metabolic rate in an experienced meditator. Contrasting determinations from similar studies of two subjects who were not yoga practitioners showed that their metabolic rates remained unchanged.

Kothari, Bordia, and Gupta (1973) reported on an Indian yogi's eight-day stay fasting in an underground chamber, during which time his oral temperature decreased from 37.2°C to 34.8°C. Young and Taylor (1998) speculated about possible reactions contributing to the suggested metabolic suppression of this and other meditating subjects.

Interactions with yogis

During travels in India, my encounters with practitioners of yoga suggested possible physiological implications of meditative episodes. A subsequent visit to a Hindu temple in Nepal provided an opportunity to witness the preparation and chamber burial of a yogi for three days for the purpose of achieving deep and prolonged meditation. Visiting friends in Kathmandu, I bicycled out each morning to observe preparations of the burial site near the prominent Pashupatinath temple. On a nearby hill a chamber had been dug measuring about one and a half meters cubed and lined with bricks and mortar. A concrete rim extended aboveground.

On the burial day the yogi, a young man clad in a simple dhoti, was ceremoniously lowered into the pit. The enclosure was then

3.2. Burial site near Pashupatinath Temple, Kathmandu, Nepal, within which the yogi remained for three days. Interior dimensions approximately 1.5 meters cubed. Concrete wall extended about 20 centimeters below ground level; interior lined with bricks and mortar. The chamber was not completely sealed against air circulation. Photo: R. Elsner.

covered with corrugated roofing and a few centimeters of earth (figure 3.2). This structure would not have been completely airtight, since its covering and the loose layer of soil doubtless allowed for limited air exchange. Also, some of the yogi's exhaled carbon dioxide would likely have been absorbed by the moist earth and fresh mortar used in its construction. Nevertheless, it seemed likely that the conditions were appropriate for challenging the yogi's metabolic endurance. Upon exhumation after three days the yogi appeared well. Later that day I talked with him, and he described the experience in terms of suppressing bodily activity in order to cleanse his mind of extraneous thought and to focus upon otherworldly contemplation.

A subsequent conversation with Craig Heller about this episode resulted in our wondering about the physiological consequences of such meditative conditions. Our speculations led to an opportunity on a subsequent visit to India to perform an experimental study near Hyderabad of a yogi in his customary daily meditation. Craig and I shared interests in aspects of comparative physiology that seemed well suited for our attention to this opportunity. Through the assistance of Indian friends we were able to recruit a suitable prospect, a devoted practitioner about thirty-five years of age.

We arranged to study the cooperative yogi by evaluating his metabolic responses during his customary early morning meditation. This would entail determining oxygen consumption and body temperatures during several hours of meditation. In an effort to provide a comfortable and unobtrusive situation for the experiment we asked our subject to chose a preferred location. Meditative composure is an essential criterion in evaluating these metabolic reactions, and it is highly likely that the meditator's responses may be adversely influenced by the unaccustomed settings and requirements of the usual laboratory environments. Therefore, a determined effort was made to accommodate minimum alteration of his customary environment for comfortable and undisturbed meditation. He suggested that an old Moghul ruin outside the city would be a suitable site for the test. We made arrangements for a power source and establishment of an improvised laboratory setting at that site.

The yogi's expired air was collected by electric pump withdrawal of air from a plastic hood positioned over his head, and its volume and gas composition were determined at frequent intervals. We tested the reliability of the experimental conditions by brief determinations of our own resting metabolic rates. Our subject was introduced to the experimental situation and became acquainted with the arrangements for data collection. Oxygen consumption was determined by open-circuit collection of his expired respirations employing a gas volume flowmeter and an oxygen analyzer. The recordings began about 3:30 in the morning, his usual time of awakening for daily meditation, which he began after a brief period of quiet rest.

The yogi remained throughout four hours in his customary cross-legged lotus sitting posture, and we confirmed that he was not asleep by direct observation and by his responses to our occasional remarks addressed to him. The beginning of meditation was followed during the subsequent hour by a decline in oxygen consumption of about 40 percent from his resting, nonmeditating rate, and this approximate value was maintained throughout the subsequent four hours. It

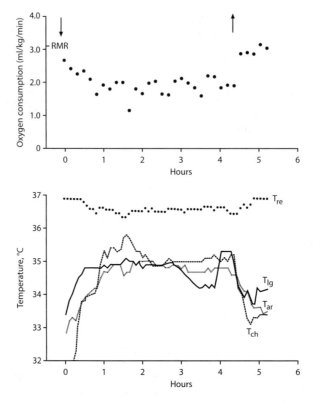

3.3.   Oxygen consumption and body temperatures of a yogi during four and one-half hours of meditation at a location near Hyderabad, India. Arrows indicate beginning and end of meditation. RMR: resting metabolic rate. *Tre*: rectal temperature. Skin temperatures: *Tlg*, leg; *Tar*, arm: *Tch*, chest. Redrawn from Heller, Elsner, and Rao 1987

represents a level of metabolism well below the reduction that might be expected to occur during sleep (figure 3.3).

The yogi's rectal temperature declined 0.5°C during meditation, compatible with his overall metabolic decline and similar to the change that might ordinarily be encountered during sleep. Skin temperatures increased moderately, suggesting that overall body heat balance remained relatively unchanged. All body temperatures resumed their previous values within thirty minutes after the termination of meditation. His oxygen consumption returned promptly to its nonmeditating level, normal for an Indian male of his age and weight on a vegetarian diet (Heller, Elsner, and Rao 1987).

We suspected that a self-induced state of reduced metabolism was brought about by what might be described as a measure of subconscious control over autonomic nervous system responses acquired through many years of practice. The result was depressed metabolism well below that recorded during human sleep. It approached that of the diving seal, clearly showing a marked lowering of metabolic activity as determined by the decline in oxygen consumption.

The yogi's increased skin temperatures and lowered rectal temperature during the four hours of meditation combine to indicate that his overall body heat content was little changed during the meditating period. His deep body and skin temperature responses resembled those of warmly sleeping human subjects. The increased skin temperature data suggests that heat loss during the meditation period might have been prevented or reduced by environmental warmth or by the experimental requirement for loosely enclosing the subject's upper body within the impermeable hood as part of the apparatus for collection of expired air. Experimental results from studies of American subjects during an eight-hour sleeping period show regular declines in both metabolism and temperatures occurring over several hours. Reduction of deep body (rectal) temperatures in those subjects averaged 1.2° C, and average skin temperatures declined about 1°C during sleep (Kreider, Buskirk, and Bass 1958).

Craig and I did another test in India; the subject was a dedicated student of yoga, an English male, twenty-nine years of age. His metabolic rate declined 32 percent during a forty-five-minute period of meditation. Results from a similar study of a twenty-six-year-old Caucasian woman who had practiced thirty minutes of meditation daily for sixteen years demonstrated a 40 percent reduction in metabolic rate during a comparable test period (Farrow and Hebert 1982).

Metabolic reactions

Thus a total of four tests of metabolic responses in experienced meditators, two Indian yogis and two Westerners, show declines of metabolism of approximately 40 percent (figure 3.4). These responses are well below the 10 percent decline that would be expected during sleep, the lowest metabolic depression ordinarily experienced by humans.

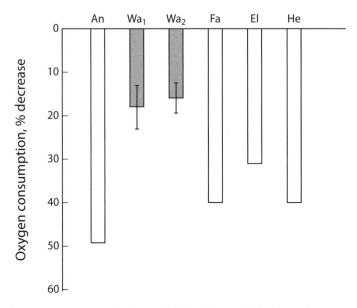

3.4. Oxygen consumption during meditation, left to right in chronological sequence of determinations. *An*: Anand, Chhina, and Singh 1961; *Wa1*: Wallace 1970 (± SD); *Wa2*: Wallace and Benson 1972 (± SD); *Fa*: Farrow and Hebert 1982; *El*: Elsner and Heller, unpublished; *He*: Heller, Elsner, and Rao 1987.

Included in the figure are similar determinations in Western students of meditation: Wa1 includes oxygen consumption data from fifteen subjects, average decline 18 percent (Wallace 1970), and Wa2, twenty subjects, average decline 16 percent (Wallace, Benson, and Wilson 1971), during thirty minutes of meditation. Their responses suggest that they routinely experience considerably greater metabolic suppression during meditation than during sleep. These, and the described reactions of highly experienced meditators, represent determinations from the literature as cited in the caption of figure 3.4.

The reduced metabolic rates in the four yogi subjects, identified as An, Fa, El, and He, an approximate 30 to 50 percent decline, were about twice those of hibernating bears and appear also to be above those of quietly diving seals (for which we do not have substantial data). They represent a nadir of healthy human metabolism. These responses during meditation suggest that experienced meditators, both yogis and Westerners, can resort to temporary depressions of resting metabolism well below those considered to be steadily maintained normal levels not otherwise subject to modification. Clearly, the statistical weakness inherent in the small number of experimental results from tests of both meditating humans and quietly diving seals suggests that tentative acceptance of this conclusion is warranted, and the need for more experimentally determined results is obvious. The techniques differed in the individual metabolic determinations, but that tends to support the combined message of metabolic reductions. These test results, considered in addition to the reported tests of students of Transcendental Meditation, represent evidence of a coherent set of reactions that are a direct consequence of the meditating condition. Further experimental clarification of the responses to meditation would be essential for what it might reveal of this condition among the repertoire of human physiological reactions. An augmented total body of evidence would be clearly necessary for adequately evaluating these responses.

Regular practice of meditation suggests that learning may influence the alterations observed in human subjects. The contrary view that responses of the autonomic nervous system are not subject to modification by learning has been dispelled by the results of numerous studies by Neal Miller and his colleagues (Miller 1969). Strictly controlled tests of experimental animals and humans revealed that the long-held view of intractability for learned responses in the autonomic system needed revision.

Physiological responses such as heart rate and peripheral circulation are subject to a multitude of influences, of which exercise and emotion are obvious examples. Therefore the isolation of autonomic reactions specifically for examination of learning effects suggests a potentially informative approach to their experimental study. Unwanted competing responses were eliminated in the animal studies by elimination of all voluntary muscle action likely to influence the autonomic reactions. That condition further supports the suggestion that modification and elaboration of the autonomic responses demonstrated by yogis and others are subject to manipulation and experimental study.

The resting metabolic rate, the condition of the quiet, nonsleeping subject, refers to the summation of the multitude of energetic reactions that combine to produce the integrated functions of the living organism at rest. It is generally more precisely defined as a basal metabolic rate in humans than in other animals, because the conditions in which they are tested, well after food consumption and exercise, can be more adequately controlled. It varies no more than a few percent in repeated determinations tested in individuals who have similar lifestyles and dietary histories. Vegetarians, likely to include most Indian subjects, characteristically have moderately lower basal metabolic rates than typical Westerners, amounting to roughly three-quarters of their values.

Such a reduced rate is evident in the study of the yogi described earlier in this chapter (Heller, Elsner, and Rao 1987), and it makes his further reduction during meditation even more unanticipated. Similar independent studies of the few examples of metabolic rate determinations, while limited in number, indicate that experienced practitioners may be deeply depressed metabolically during sustained meditation. They suggest that humans are capable, in appropriate conditions and with training, of achieving a considerable extension of metabolic range including its depression to unusually low values, even to a level twice that of the hibernating bear.

Yogis and seals, breath-holders

An unexpected observation has been made of some well-practiced meditators suggesting that an alteration of breathing frequency and depth is a likely candidate as an important factor in the initiation of respiratory depression during meditative practice. It is the occurrence of repeated breath-holding episodes during meditation. Farrow and Hebert (1982) noted similar intermittent suspensions of breathing during meditation in a group of Westerners who had practiced Transcendental Meditation for several years. Periods of breath-holding lasted from approximately ten to sixty seconds and were generally not followed by compensatory increase of breathing rate or depth. Decreased sensitivity to increased respiratory carbon dioxide has been suggested as a consequence of this specific breathing technique as it is practiced by some meditators (Stănescu et al. 1981). Its occurrence during meditation has been reported in the review by Jevning, Wallace, and Beidebach (1992) and by Kesterson and Clinch (1989).

Some other observations attest to the possible importance of this respiratory manipulation in the reactions of the meditator. A tolerance of elevated concentrations of carbon dioxide that would ordinarily be sufficient to stimulate increased breathing was reported in the Anand, Chhina, and Singh (1961) study of a dedicated yogi. During the test, oxygen concentration within the closed box in which it was tested declined from the 21 percent level in air to 15 percent in accordance with its consumption by the yogi. His exhaled carbon dioxide increased from its negligible value in air at the onset of the test to more than 5 percent, a concentration ordinarily capable of distinctly noticeable stimulation of increased respiratory drive. However, the yogi did not show the typical response of increased rate and depth of breathing consequent to the increasing carbon dioxide concentration.

Contrasting results were obtained from tests of the two control subjects, whose respirations steadily increased as a result of stimulation by the increased carbon dioxide levels. We (Heller, Elsner, and Rao 1987) have suggested that this tolerance of elevated environmental carbon dioxide resembles the breath-holding seal's or the hibernating mammal's during the onset of torpor (Malan 1988; Bickler 1984).

Breath-holding may have contributed to the reactions determined in the subject of our study (Heller, Elsner, and Rao 1987). During discussions after his testing, the yogi subject described for us a procedure that he employs for supporting breath suspension. It involved the displacement of his tongue toward the back of his mouth to a position likely to prevent or inhibit the flow of normal respiration. This maneuver had been made possible years earlier by his making incremental small cuts over several days on the lingual frenulum, the membranous tissue extending from beneath the midline structure of the tongue to the floor of his mouth, thus freeing the tongue for its rearward displacement. He reported that this manipulation was a not uncommon practice among dedicated Indian meditators of his acquaintance. Frequent use of this practice is likely to result in diminished responsiveness to increased carbon dioxide, such as was noted in the yogi subject of the Anand, Chhina, and Singh (1961) study. The lack of response to the elevated carbon dioxide content of air in the yogi's environment contrasts with the vigorously increased respiratory reactions of the control subjects of this experimental series.

Anecdotal reports of the regular practice of frequently repeated breath-holding during yogic meditations suggests that it may be routinely practiced and may have some suspected but undocumented effects on the yogi's metabolism. Such a practice, if sufficiently lengthy and challenging, may be expected to result in a gradual decrease of the ordinary respiratory stimulation from increased carbon dioxide that would be experienced during breath-holding (Farrow and Hebert 1982). It resembles the suppression of the normal vigorous respiratory reaction to inhaled carbon dioxide that develops from repeated exposure, as observed in some breath-holding Japanese ama divers (Masuda et al. 1982) and in those trained for escape from disabled submarines (Schaefer 1965). A similar retention of carbon dioxide occurs in animals during the onset of hibernation (Malan 1988; Snapp and Heller 1981).

The frequency of this technique's use for meditative enhancement and of repeated breath-holding as a general accompaniment to meditation is a matter of conjecture. It may be a regular occurrence by dedicated practitioners, as suggested by the use of frequent breath-holding by participants in our study. While the practice of tongue displacement may provide a ready blockade or hindrance of normal respiratory flow, it is perhaps not expected to be an essential component of respiratory suppression during meditation. It is more likely to be one technique among several that are employed for development of the most effective contemplative engagement. The physiological consequences of its regular use would be anticipated to include gradual suppression of the carbon dioxide respiratory stimulus.

The yogi of the Anand, Chhina, and Singh (1961) study was notably tolerant of increasing carbon dioxide, which would ordinarily be expected to increase the rate and depth of breathing, suggesting an adjustment to his frequent breath-holding bouts. Such reduced respiratory sensitivity to elevated carbon dioxide resembles the tolerance of increased respiratory carbon dioxide in seals. This and perhaps other manipulations of respiratory excursion that are likely to result in unusually high carbon dioxide retention have been noted in some other yoga practices. These involve deep inhalations and expirations alternating with breath-holding periods. Such respiratory manipulations can be expected over time to result in diminished responsiveness to the customary respiratory stimulation of increased respiratory carbon dioxide, as was noted in the study of Anand, Chhina, and Singh (1961). The resulting carbon dioxide retention can be expected to induce generalized modest cellular acidosis, and this condition may contribute to metabolic suppression during meditation. Frequent repetition of this practice is likely to substantially reduce the usual vigorous response to elevated respiratory carbon dioxide noted in subjects who have not undergone such a respiratory manipulation and its long-term effect.

The respiratory and metabolic implications of apparently intentional suppression of increased breathing by carbon dioxide stimulation have not been examined, and its possible consequences are an obvious topic for critical study. The dulled response to increased carbon dioxide in the meditating yogi resembles the similarly suppressed reaction of the diving seal. It suggests that they may employ mechanisms governing this condition that are comparable in some respects. A possible consequence of the increase in carbon dioxide retention is its contribution to a reaction resembling that resulting from the protective effect of cardiovascular preconditioning (Zong et al. 2004), considered in a later chapter. Such an effect may be a contributing factor in the development of the experienced yogis' metabolic suppression during meditation, a possible and testable connection relating the reactions of meditating yogis and diving seals.

CARDIOVASCULAR AND

METABOLIC INTERACTIONS

IN DIVING SEALS

An essential component of the seal's diving capabilities, its cardiovascular responses, has been mentioned briefly in the preceding chapters. Its fundamental contribution to the seal's diving endurance has been recognized and its early study by Irving and Scholander has been reviewed in chapter 2. Its subsequent examination by others who have tested these reactions added further insights into how the submerged seal's capabilities are implemented (reviews: Andersen 1966; Elsner and Gooden 1983; Kooyman 1989).

The regular functional role clearly attributable to the mammalian circulation is the maintenance of blood flow to tissues in response to their needs for nutrients and oxygen and for the removal of metabolic end products. Accordingly, sustained metabolism of an organ or tissue depends upon its circulatory supply. The cardiovascular requirements supporting metabolic activity depend upon adequate circulatory support and its oxygen-based metabolism. Continued activity subsequent to depletion of readily available oxygen depends on anaerobic metabolism, amounting to a small fraction of the comparable output that may be realized from oxidative processing of the same resource. The seal's abundant tissue glycogen can provide fuel for the continuing anaerobic metabolism that accompanies these responses during long dives (Kerem, Hammond, and Elsner 1973). Cellular function finally ceases as metabolic end products accumulate and the supporting glycogen resource declines, a condition that imposes a limit on breath-holding and dive duration.

A recurring theme in reports of marine mammal diving adaptations supports the view that reactions primarily responsible for protecting the integrity of the central nervous system are expressed in the cardiovascular system. These effects can be recorded in free-diving seals, as they have been in several species in which they have been studied. Responses of the heart and circulation feature prominently in the protective reactions of marine mammals during long dives. Reduced cardiac output and constricted peripheral blood flow impose major alterations on the circulatory system. Their combined reactions result in lowering of overall oxygen consumption by reducing blood flow, sometimes apparently eliminating it in kidneys, intestines, skin, and other organs that can better tolerate temporary circulatory arrest. The continuing blood-pumping effort of the heart is also reduced in accordance with the lowered requirement for normal nondiving circulation. These responses are, in turn, much influenced by whatever competing demands there may be for circulatory support of other activity, especially that required for muscular exercise. Thus the maintenance and extent of reduced blood flow distribution are variable depending upon several aspects of the diving condition and its duration. Muscular exercise in support of swimming, for example, is likely to require a substantial fraction of available blood flow.

The reduced blood flow distribution during dives results in an abrupt decline of metabolism in the blood-deprived organs. An obvious example is demonstrated when circulation in an arm or leg is prevented by a tourniquet, thus depressing metabolic activity in the downstream tissues. Along with the diving seal's reduced circulation, its heart rate slows, sometimes drastically to no more than a few beats per minute. This bradycardia is initiated by neural reflex activation and occurs in coordination with the lessened demand on the heart's pumping action for supplying the consequently reduced output of blood. The demand for cardiovascular supporting reactions is variable depending on whether the diving seal is swimming or quietly resting.

These reactions were originally observed in controlled laboratory experiments, and it was subsequently determined that similar responses occur in free-diving seals. The uterus of the pregnant seal is an exception; it is supplied by continued perfusion during the dive (Elsner, Hammond, and Parker 1970). The original studies of seal diving by Scholander and Irving, reported in preceding pages, have been extended to a variety of species and conditions (Elsner et al. 1966; Elsner and Gooden 1983; Elsner 1999; Kooyman 1989; Thompson and Fedak 1993; Zapol 1996; Butler and Jones 1997; Ponganis, Kooyman, and Ridgway 2003; Ramirez, Folkow, and Blix 2007).

The marked depressions of heart rate noted in figure 2.2 were recorded from ringed seals during unrestrained excursions swimming slowly for a few minutes while submerged under ice. This bradycardia contrasts with the unchanged or modest slowing of heart rate of relaxed dives near a breathing hole (Elsner et al. 1989). Forced dives usually produce vigorous responses, as might be expected, due to the imposition of restraint. They resemble the reaction to lengthy dives of unrestrained seals, the animals responding to both conditions by enhanced activation of mechanisms for endurance of long or exercising dives.

Metabolic reactions in dives

Experiments undertaken by Scholander, Irving, and their colleagues in the 1930s and 1940s demonstrated the generally elevated metabolic rate of resting, nondiving seals, contrasting sometimes with that of comparably sized terrestrial mammals. Scholander's 1940 work also showed that the nonexercising metabolic rate of the quietly submerged seal equaled about one-half or less of its comparable nondiving rate. The reduction of the seal's metabolism while submerged is a likely consequence in large part of the lowered blood flow distribution within those portions of its body that can tolerate withdrawal from activity for the duration of the dive. This circulatory redistribution thus provides necessary support for the central nervous system and other minimum essential metabolism.

Results of those early studies have been extended in various ways by subsequent investigating procedures, and the reactions of marine mammals in their aquatic habitats continue to attract considerable interest. There has been some controversy regarding the nature of tests intended to compare metabolic responses of marine and terrestrial species. The varying results suggest that the seals being tested may not be in a truly resting, postabsorptive, nondiving state for such determinations. An experiment designed to meet these conditions was performed with California sea lions that were trained to submerge without exercise. Results indicate a resting metabolic rate of two to three times that of similarly treated terrestrial mammals (Hurley and Costa 2001). These tests were done with unrestrained animals that had been thoroughly conditioned for becoming accustomed to the experimental arrangement. Sea lions and seals are highly trainable animals, and their understandably adverse reactions to forced experimental dives can be made less traumatic by their becoming trained to the procedures and accustomed to the persons conducting them.

The seal's enhanced oxidative metabolic capacity is supported by the abundant oxygen resource of its circulating hemoglobin and its copious blood volume. Additionally, the myoglobin content within its muscle cells, several times higher than in terrestrial mammals, as described in chapter 2 (data from several sources are summarized in Kanatous et al. 1999 and Polasek, Dickson, and Davis 2006). The enhanced reserves of hemoglobin and myoglobin combine to provide the diving seal's requirements for oxygen-dependent metabolism. Appropriate use of this supporting resource depends on the animal's facility for assuring its ready and sustained use when needed during long dives. Research by Watson et al. (2007) has shed light on how the seal's oxidative reserve can provide continuing metabolic activity even at low oxygen partial pressure, such as occurs in lengthy dives. The evidence shows that the oxygenated myoglobin is located in close proximity to mitochondria within the seal's muscle cells. This localization is likely to enhance diffusion and thereby maintain metabolic activity as oxygen pressure falls during lengthening dives.

Seal species have been observed to differ in their activity during dives. Their levels of exercise, foraging that involves vigorous swimming contrasted with quiet rest, differ among species, and their cardiovascular reactions vary accordingly (Andrews et al. 1997). Species differences in submerged activity and in physiological responses likely depend upon relative metabolic demands, capacities for oxygen blood and tissue contents, and dependence on anaerobic metabolic resource while submerged (Andrews et al. 1997). Gray seals make frequent quiet, nonexercising dives with related reduced metabolic cost (Thompson and Fedak 1993); elephant seals engage in repeated long dives and only brief recovery intervals (Le Boeuf et al. 1988); Weddell seals appear to require lengthy surface recovery from dives (Kooyman 1989).

Cessation of blood flow, anaerobic metabolism

An aspect of the diving seal's metabolism is pertinent to this discussion. It relates to that portion of the seal's total metabolic activity that is supported by nonoxidative processes, the anaerobic component. Its contribution depends upon the energetic resource derived from conversion of stored tissue glycogen to lactic acid. This metabolic reserve results in obligatory accumulation of reaction products, specifically lactic acid and hydrogen ions. Despite the relative inefficiency of this dependence on anaerobic glycolysis and the requirement for neutralization of the acidic reaction products, this resource serves the important function of sustaining metabolic activity during intervals when adequate oxygen reserves have been temporarily reduced or exhausted. Seals are unusually well equipped for exploitation of this capacity by reason of their considerable tissue glycogen stores (Kerem, Hammond, and Elsner 1973), which are readily convertible into the metabolic resource of glucose (figure 4.1).

The magnitude of the glycogen reserve in seals supports their unusual capability for sustaining anaerobic exercise further beyond the maximum level of oxygen-based activity. The contribution of this unusually high reserve has been demonstrated in tests of the seal's exercise capacity (Ashwell-Erickson and Elsner 1980). In the response to exercise of humans and land mammals, augmented oxygen consumption

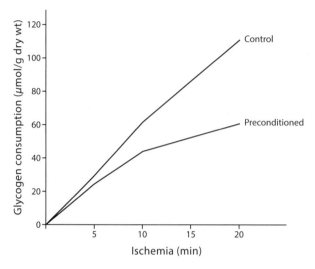

4.1. Glycogen consumption after elimination of blood flow, preconditioned and control, demonstrating reduced metabolism of preconditioned dog heart. Redrawn from Murry et al. 1990.

associated with increasing workload reaches a plateau value identifying the maximum metabolic capacity, an indicator of exercise limit for performing work. Efforts to continue increasing the workload beyond that level are abruptly curtailed by exhaustion, approaching maximum demand upon metabolic resources. Seals, however, when tested for their response to exercise in an "aquatic treadmill," showed an enhanced ability for drawing upon anaerobic metabolic reserves in support of continued activity (Elsner 1986).

The tolerance of long diving by seals is similarly enhanced by increased capacity for anaerobic metabolism based on considerably augmented glycolytic reserves and related enzyme activities. Discussion of the problem and related questions appears in a review by Castellini (1985), the summary of which includes these words: "The tissues probably survive hypoperfused periods by both depressing metabolism and activating glycolysis. In such a scenario, metabolic depression would account for most of the survival ability (a reduction of potentially 90%) and glycolytic ATP production would eliminate the energy deficit."

Further examination of this issue is considered in a report of metabolic rates in diving Weddell seals (Castellini, Kooyman, and Ponganis 1992). The rate of oxygen consumption during the longest dives was found to be reduced to values below those of shorter quiet dives or sleep. The seal's declining metabolism in the longest dives suggests a generally lowered metabolic cost of submersion, a value additionally increased by the accompanying reduced level of muscular activity. Lengthy dives are usually those involving little exercise.

Determination of metabolic costs of submersion in a seal during the quiet, nonexercising condition have been difficult to obtain. Uncertainty regarding that essential information regarding the seal's state during the dive is accounted for by the difficulties of determining the nature of the activity, or lack thereof, by the submerged animal. Estimates of the likely range of metabolic costs of the nonexercising dive were made by Hochachka (1986), giving a value of 25 to 45 percent of the resting non-diving metabolic rate.

The seal's reliance on elevated capacity for anaerobic metabolism is demonstrable in its sustained cardiac effort during long dives. Contrasting with the increased cardiac rate and metabolic effort of exhausting exercise in terrestrial mammals, the pumping action of the seal's heart is maintained despite curtailed oxidative metabolism, depending partly on anaerobic reserves represented by its abundant glycogen stores (Kerem, Hammond, and Elsner 1973; Elsner et al. 1985). This capability for extended anaerobic support of cardiac function is unusual in mammalian hearts, which are more generally subject to impaired operation due to early exhaustion and cessation of function when subjected to reliance upon anaerobic metabolism.

The diving performance of other marine mammal species suggests considerable variability in exercise capability along with variations in breath-holding endurance. The routine dives of dolphins, for example, indicate that their capacities generally limit them to shorter dive durations than those of seals. Their brief dive endurance is related to their relatively modest oxidative and anaerobic reserves (Williams, Haun, and Friedl 1999). The effort of underwater propulsion is considerably reduced by intermittent gliding behavior sometimes also observed as a regular mode of propulsion in Weddell and elephant seals, bottlenose dolphins, and blue whales (Williams et al. 2000; Williams 2001).

The seal kidney

Tolerance of the seal's kidneys for temporary deprivation of sustenance from blood flow has been verified in several studies (Halasz et al. 1974; Hong et al. 1982; Hong 1989). An experiment designed to test the notion that seal kidneys differ from those of their terrestrial cousins demonstrates their superior tolerance of long circulatory arrest without subsequent disturbed function. The purpose was to compare the effects of arrested circulation in seal and dog kidneys. A study of kidneys isolated from the donor animals and subjected to discrete periods of no blood flow permitted direct testing of comparative responses and survivability. The experiment consisted of testing isolated seal and dog kidneys in an experimental arrangement that provided perfusion at constant pressure with oxygenated blood previously collected from the animal and allowed regular sampling of urine production and renal function tests. The kidneys were then deprived of perfusion for one hour, and the extent of functional recovery of kidney function was examined during subsequent restoration of blood flow (Halasz et al. 1974).

The experimental results showed a considerable difference between kidneys of the two species in tolerating the cessation of blood perfusion. Dog kidneys were severely depressed after identical treatment, blood flow remaining much reduced during the recovery hour, and they showed drastically reduced production of urine. In contrast, seal kidneys' blood flow and function were rapidly restored after the hour-long ischemia, and they produced urine at rates equal to those before the experimental period. Seal kidney oxygen consumption recovered well, in marked contrast to the nonfunctioning reaction of dog kidneys (figure 4.2). Seal kidney functions were further evaluated by tests of ion transport and found to be unaffected by the conditions of hypoxia and acidity, while those of similarly tested rat kidneys were severely disrupted (Hong et al. 1982).

The conclusion from these studies indicates that the seal kidney may tolerate lengthy reduction or deprivation of circulation that would depress normal function in the kidneys of dogs and likely that of other terrestrial mammals. This tolerance probably depends upon the seal kidney's anaerobic metabolic facility for maintaining continued activity when challenged by deprivation of blood flow. Such

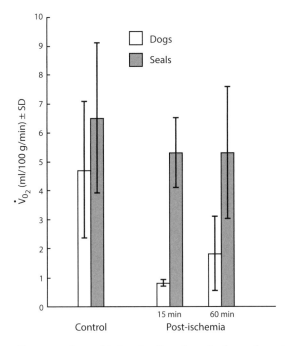

4.2. Recovery from experimental ischemia of one hour in dog and spotted seal (*Phoca largha*) kidneys. Redrawn from Halasz et al. 1974.

a condition applies as well to the resistance of other seal organs to similar ischemia. Depression of metabolism during dives is a likely consequence of that condition. Scholander (1940) showed that seals recovering from experimental dives did not fully compensate for the "oxygen debt" imposed during the dive. The deficit amounted to about one-half of the metabolic activity that would have sustained the nondiving animal. A possible contributing factor may be that the seal's enhanced anaerobic capacity, despite its augmented condition, is insufficient to fully compensate for the deficiency of metabolic activity imposed during long dives. The effect resembles a similar condition relating to the absence of or modest increase of oxygen debt that has been detected in newborn mammals following a depression of metabolism resulting from hypoxia (Rohlicek et al. 1998).

The metabolic resources within the bodies of some marine mammals are augmented by their increased glycolytic reserves in the form of tissue glycogen, notably elevated in the critical brain and heart

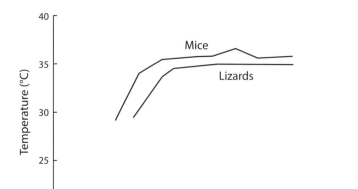

4.3.   Effects of reducing oxygen on temperature of mice and lizards. Redrawn from Wood and Gonzales 1996.

tissues of seals (Kerem, Hammond, and Elsner 1973), and their resistance to cellular hypoxia (Kerem and Elsner 1973a). An additional hazard is incurred by extended interruption of breathing that takes place when respiration and circulation are restored after a period of asphyxia or ischemia. Reperfusion with oxygen-rich blood after circulatory interruption is a potential source of cell damage as a consequence of the ensuing *oxidative stress*, a condition related to the generation of superoxide radicals and other reactive oxygen species that in some circumstances may be harmful to cardiac metabolic processes. This potential threat arises from the possibility that the natural enzymatic defenses may be overwhelmed in these conditions. Temporary maintenance of cardiac tissue acidity appears to prevent or moderate these adverse effects (Kitakaze et al. 1997).

The effects of decreased oxygen availability on a related lowering of body temperature have been described in several studies. This reaction to hypoxia results in induced hypothermia in a variety of animal species ranging from protozoa to mammals (Wood and Gonzales 1996; figure 4.3). The mammalian response is especially notable in the associated tolerance of infant animals to both low oxygen and cold exposure, which generally exceeds that of adult animals of the same species. Similar lowering of deep body temperature by a few degrees occurs frequently in diving seals (Scholander 1940).

Oxygen sensing and free radicals

Multicellular organisms depend upon a steady source of oxygen for maintaining their metabolic activity. Disruption of that supply is tolerated for limited durations that vary with their capacities for anaerobic metabolism. Respiratory mechanisms range among species from simple diffusion in primitive organisms to the elaboration of circulating blood in vertebrate species. The maintenance of oxygen homeostasis is a primary organizing principle of mammalian life (Hochachka and Lutz 2001).

Circulation of blood and its function in gas transport is established early in fetal life, and the matching of this supply to the needs of the fetus is essential for its growth and health. The formation of new blood vessels is required during fetal growth to support this developing respiratory function. The fetus survives at the downstream terminal of maternal oxygen supply, thus rendering it exposed to low blood-borne oxygen values.

Oxygen, the essential substance upon which aerobic metabolism depends, is stored for moment-to-moment accessibility in blood hemoglobin and muscle myoglobin. The several times greater oxygen storage capacity in these critical regions of the seal's body than in nondiving mammals, noted before, is a resource upon which the seal depends for lengthy dives. The steady decline of blood and tissue oxygen content during dives leads inexorably to cellular hypoxia (along with increased carbon dioxide and lactate), its severity depending on the length and conditions of the dive. Among the mechanisms for dealing with exposure to this oxygen deprivation are expressions of genetic effects specifically resulting in the elimination or moderation of that exposure. The requirement for oxygen homeostasis is related to the maintenance of an ideal condition in which neither cellular hypoxia nor the toxic effects of excessive oxygen levels are realized.

Mammalian well-being and ultimate survival are intimately associated with ready access to oxygen, and it is therefore not surprising that animals have a well-developed capacity for sensing the oxygen levels in various tissues, depending upon their individual requirements. Arterial oxygen tension is detected in strategically positioned specialized structures in the arterial circulation, the chemoreceptor carotid and aortic bodies. The monitoring and correction of disturbed oxygen levels include both central and local sensing mechanisms. Cellular sensing of hypoxia is generally followed by a decline in energetic activity associated with conversion to the less efficient anaerobic metabolic mode. This condition leads to expression of specific genes relating to glycolysis and to the induction of the hormone erythropoietin (EPO) genes. Augmented production of EPO supports hemoglobin production, thus responding to increased need for oxygen transport (Bunn and Poyton 1996).

The hypoxia-inducible transcription factor HIF-1 is present in cells and activated by exposure to hypoxia (Semenza 2000). It is one of several factors that respond to decreases in available oxygen, and it is responsible for initiating the regulation of long-term genetic responses to hypoxia. It is also associated with an important genetic factor for modulating tissue blood supply, vascular endothelial growth factor (VEGF).

Oxygen radicals, the reactive oxygen forms that are implicated in the origin of oxidative stress, are in lesser concentrations essential biological messengers for the signaling of required cellular oxygen tension adjustments (Kietzmann, Fandrey, and Acker 2000; Hermes-Lima and Zenteno-Savín 2002). This effect suggests a possible connection worthy of special attention as applied to diving animals. Analysis of ringed seal tissues showed the presence of HIF-1 in several organs, highest in skeletal muscle and lung (Johnson, Elsner, and Zenteno-Savín 2004, 2005). It indicates a response that invokes the HIF-1 system in anticipation of the requirement for avoiding the negative effects of alternating ischemia and reperfusion that are associated with frequent dives.

Another source of hypoxic exposure is high altitude. A comparison of the physiological reactions to that environment with those of the breath-holding diver may help to clarify the comparative responses of these differing exposures to hypoxia. The regular decline of atmospheric oxygen and the accompanying decrease in lung alveolar and arterial gas pressures that accompany ascent to high altitude stimulate respiration. The decline in blood oxygen is sensed by specialized cells within the carotid bodies that are perfused by arterial blood on its route to the cerebral circulation, resulting in increased breathing and leading to increased lung and blood oxygen, lowered carbon dioxide, and more alkaline blood. The acclimatization process and related improved respiratory responses in mammals require days to weeks for development.

Breath-hold diving also produces a respiratory stimulus, but one that is suppressed during the dive, preventing inhalation of water. The hypoxia, elevated CO_2, and increasing acidity consequent to breath-holding are the conditions defined as asphyxia. This combination can result in extreme values of lowered oxygen, increased carbon dioxide, and increased acidity—the hypoxia-hypercapnia-acidosis combination—as the seal's long dive progresses. This combined respiratory condition is initiated at the onset of breath-holding at the start of the dive. Superior endurance of these conditions characterizes the seal's adaptation for long submergence. Its reduced oxygen requirements during the dive are satisfied by withdrawals from storage in lungs and in circulating hemoglobin, a resource unusually elevated in many marine mammal species.

The seal's tolerance of diving hypoxia suggests that it might be unusually tolerant of exposure to the low oxygen of high altitude, a condition it is highly unlikely to encounter in its ordinary lifestyle. A testing of that possibility was done in which seals were taken to ten thousand feet in the California Sierra and maintained there for four weeks. They showed evidence of immediate and successful respiratory adjustment to this highly unusual form of hypoxic exposure (Kodama, Elsner, and Pace 1977). Thus, it suggests that a diving lifestyle and high-altitude living have some similarities relating to physiological adaptation, both reacting to challenging respiratory inadequacies in those extreme environments. Such an interpretation might be inferred from a consideration of similarly increased circulating red cells (elevated hematocrit) in both diving mammals and terrestrial mammals adapted to high-altitude hypoxia. Both of those environmental adaptations may, however, be limited by the the complicating feature of the inevitable increase in blood viscosity that results from increased production of red blood cells.

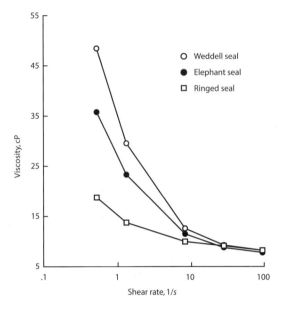

4.4. Viscosity of blood (centipoise) in Weddell seal (*Leptonychotes weddelli*), northern elephant seal (*Mirounga angustirostris*), and ringed seal (*Phoca hispida*). Wickham et al. 1990; Meiselman, Castellini, and Elsner 1992.

Resistance to blood flow and the resulting increased cardiac load are the inevitable consequences of elevated blood viscosity. Weddell seals, and some other seal species, avoid this potential difficulty by sequestration of red cell deposits in the spleen and venous reservoirs when not routinely diving (Qvist et al. 1986). Elephant seals may use a similar technique, but it seems clear that ringed seals, comparatively brief divers, maintain relatively low hematocrits throughout (Wickham et al. 1990; Meiselman, Castellini, and Elsner 1992). The high hematocrits and related increased blood viscosities are illustrated in figure 4.4. Human blood viscosity lies roughly midway between the elephant seal and ringed seal values.

A similar inevitable increase of blood flow resistance from increased viscosity is a consequence of augmented hemoglobin and red cell production above that required for optimum oxygen transport at high altitude (Garruto and Dutt 1983; Beall 2007). Hemoglobin concentrations in the range of 15 to 18 grams per 100 ml of blood provide maximum oxygen transport, and that function can be expected to be reduced at both lower and higher concentrations (Villafuerte, Cardenas, and Monge-C. 2004). Garruto and Dutt (1983) point out that much of the historic blood data from Peruvian high-altitude natives has been derived from mine workers and relatively recent migrants to the region from lower altitudes, both groups likely subject to relatively recent augmentation of blood values and related blood viscosity. Determinations of red blood cell and hemoglobin levels of inhabitants of Andean mining communities were found to be notably higher: 25 to 30 percent compared with 10 to 12 percent in traditional farming populations living at similar altitudes.

The discovery of the important role played by the unexpected influence of nitric oxide gas inhalation on pulmonary circulation and oxygen transport to peripheral tissues has provided explanation for diverse cardiopulmonary reactions. Increases in oxygen delivery via elevated pulmonary circulation without increasing pulmonary arterial pressure are thus indicated to compensate for the reduction in ambient oxygen of high-altitude Tibetans (Hoit et al. 2005). The interpretation of an appropriate optimum hematocrit range for blood gas transport has been derived from studies of humans at high altitude, but possible similar effects of circulating red cells in marine mammal blood have not been identified. Elevated blood viscosity and its likely contribution to increased resistance in peripheral circulation might be speculated in some circumstances to impose a limit on the diving seal's performance, especially when the demand for oxygen consumption is increased during exercise.

Temperature and hypoxia

Exposures to hypoxia and low temperature are linked in a combination of operating adaptive mechanisms. Both can elicit responses of reduced metabolism. That is, the combined exposure to hypoxia and hypothermia can result in a metabolic alteration into a protective condition of declining oxygen demand. The usual mammalian response to hypoxia is a prompt compensatory increased rate and depth of breathing and an eventual recourse to anaerobic metabolism. Anaerobic resources, however, are constrained by the lesser ATP production from glycolysis. Furthermore, acid products of glycolysis require detoxification, and exceeding the limited capacities for that process quickly inhibits the metabolic machinery and brings it to a halt. Longer-term reactions (days to weeks) to frequent or chronic hypoxic exposures include gene expression of the earlier mentioned hypoxia-inducible factor 1 (HIF-1), which augments glycolytic activity and stimulates red cell synthesis and capillary growth (Wenger 2000).

The seal's ability to tolerate the respiratory challenge of long dives may depend in part on a decrease in body temperature of a few degrees C, depending upon species and the submerged duration. The diving animal's ability to suppress its metabolism and to realize thereby the temporary benefit of reduced need for steadily maintained respiration is intimately connected with its regulated body temperature, cooling associated with lowering of metabolic activity. Such responses have been noted in several studies of diving seals (Scholander, Irving, and Grinnell 1942b; Hammel et al. 1977; Kooyman et al. 1980; Hill et al. 1987). The decrease in body temperature during long dives may be more than 2°C in some species. A fall of several degrees C has been observed in diving elephant seals. An accompanying modest rise in skin temperatures would suggest that total body heat loss during dives may be hastened by heat transfer from skin to water. However, the transient nature of temperature changes in long-diving adult elephant seals may have little effect on overall heat balance (Meir and Ponganis 2010).

Sleeping and diving

There have been numerous reported observations of sleeping seals that appear to mimic in some respects the reactions that they show during diving. These are particularly noticeable in immature seals resting on land before the beginning of their ventures into the sea. Postweaning elephant seal pups have been observed routinely breath-holding for several minutes while appearing to be asleep on the shore. The sleep pattern in elephant seal pups reveals intermittent periods of brief breathing followed by several minutes of breath-holding. Adults of the same species regularly breath-hold for many minutes while resting on land, some as long as twenty-five minutes (Andrews et al. 1997). Similar episodes can be seen to occur in captive adult elephant seals while submerged. A pattern of quiet sleep while submerged on the bottom of the tank is interrupted by slow ascent to the surface, a few breaths, and return to the bottom, all the while still appearing to be asleep. These episodes have sometimes been observed in zoos, alarming visitors concerned that the animals may be lying dead on the tank bottom.

Sleep apnea, breath-holding during sleep, was described in gray seals by Ridgway, Carter, and Clark (1975), and it has since been documented in several seal species. Some speculation has been directed to its possible significance. One possible explanation relates to the animal's need for sleep while venturing far at sea. Elephant seals, for example, may remain at sea for several months and engage in repeated diving periods with brief intervals of breathing at the surface. Sleep patterns recorded in brain wave recordings (electroencephalograms, or EEGs), reveal two dominant characteristics: slow-wave sleep (SWS) and rapid eye movement (REM) rhythms. These patterns are apparent in recordings from sleeping seals, showing that they conform with minor variations to similar patterns in other sleeping mammals. There are, however, some variations among species in breathing patterns and recorded EEGs during dives. These studies have confirmed that sleep can continue during respiratory cycles of dives and surface breathing (Castellini 1996).

Considerations of the diving seal's special tolerance for long submergence would be incomplete and overly simplified without concern for how its reactions are initiated and governed. Seals' apparent ease and grace while submerged contrast with our far more restricted and awkward underwater efforts. But we both depend upon regulatory mechanisms that operate similarly in their essential functions; their expression for the demands associated with breath-holding is obviously much more developed in seals. Frequent dives demand exquisite timing of mechanisms that regulate respiration and circulation; long sustained immersions result in metabolic depression. Some aspects of their organization have been examined, and what we know of this aspect of the seal's life shows it to be a well-tuned extension of comparable general mammalian reactions.

In comparison, we lack knowledge of what regulates the meditating yogi's metabolic depression, and we are left with speculations based on suggestions derived from few and inconclusive observations. Testable inferences may, however, come from what is suggested by the diverse and provocatively similar reactions among numerous species, including the diving seal's depressed metabolic responses. The vital functions of mammalian life concerned with oxygen consumption, carbon dioxide elimination, and acid–base balance are monitored by components of the central nervous system, integrated and responded to by associated reflexes. The mechanisms concerned with their working details in terrestrial mammals have been subjects of long and vigorous research, but the modifications of these reflexes for functioning in diving seals have been examined relatively recently. Many questions remain, but the outlines of regulatory pathways are becoming clear. Brief remarks that appear in previous pages regarding these reactions may be usefully amplified here.

The seal's heart

Responses of the seal's heart, among its more vulnerable organs, and circulatory functions during dives represent an extension of the regular protective mechanisms that exist in other mammalian hearts. Healthy cardiac function depends upon an uninterrupted source of oxygen and steady removal of the metabolic residue, carbon dioxide and acid products. The heart's coronary circulation serves these functions, and its blood flow responds to the varying needs of the cardiac muscle as the heart rate and blood-pumping effort vary. Oxygen consumption of the diving seal's cardiac muscle, supplied by the coronary blood flow, is reduced as heart rate slows and cardiac output is decreased to match the lessened need for perfusion of peripheral tissues (Kjekshus et al. 1982). Continued functioning of the seal's heart when its blood-borne oxygen decreases as the dive lengthens depends on its enhanced capacity for reliance upon anaerobic energy sources, a capability generally greater than that of nondiving mammals. Increasing cardiac production of lactate that occurs as the dive progresses is a product of the conversion to anaerobic metabolic processes and is supported by the seal heart muscle's unusually high glycogen content, a resource exceeding that of terrestrial mammal (dog or pig) hearts (Kerem, Hammond, and Elsner 1973).

Despite the advantageous metabolic nature of the seal's heart for enduring long dives, it resembles other mammalian hearts in that it does not tolerate well total deprivation of nutritive supply in cardiac muscle. Experimental occlusion of local coronary blood flow in the seal heart results in prompt failure of myocardial contraction in the related discrete downstream region. It is a clear indication of acute local myocardial dysfunction resulting from the decline of energetic resource and the accumulation of resulting metabolic products (Elsner et al. 1985). Removal of these substances by intermittently maintained coronary blood flow, even at lowered rates of perfusion, is essential for sustained function, along with a steadily renewed supply of oxygenated blood.

Seal coronary blood flow varies regularly along with the slower heart rate during dives, sometimes ceasing for a few seconds to a minute before restoration of flow (Kjekshus et al. 1982). Such unusual behavior of alternating myocardial circulation and arrested blood flow is consistent with maintained cardiac metabolism that alternates regularly between oxygen-supported and anaerobic energy sources. This process is sustained by the seal heart muscle's high glycogen content (Kerem, Hammond, and Elsner 1973), a resource for fueling anaerobic metabolism. Its intermittent coronary blood flow repeatedly renews the supply of oxygen-rich blood and removes accumulated end products, thus making available the full complement of the cardiac energetic resources.

The terrestrial mammal's heart, in contrast, is more dependent on oxidative energy sources, and when its blood flow is reduced its ability to continue normal function on anaerobic reserves is severely limited. Despite its intolerance of abrupt total deprivation of blood flow, the seal's heart tolerates exposure to the increased metabolic products of long dives. But this protection ultimately depends upon the circulatory removal of these substances. Thus, cardiac function continues during dives under conditions of reduced oxygenation if the intermittent blood flow removal of metabolic by-products, such as carbon dioxide, lactate, and hydrogen ions, can maintain their myocardial concentrations below levels that may inhibit the heart's operation.

When anesthetized seals with intact coronary circulations were subjected to decreased oxygen supply, their hearts continued to maintain functional integrity longer (seventeen minutes) than did pig hearts (eight minutes). Furthermore, the seal heart promptly recovered normal function when reoxygenated; none of the similarly treated pig hearts survived this treatment (White et al. 1990). Histological examination of the hypoxia-exposed pig hearts revealed extensive structural damage in contrast to the nearly intact condition of the seal heart muscle, consistent with its capacity for operating anaerobically when required to do so. The experimental contrast with impaired function of the hypoxic pig's heart indicates the seal's superior ability, dependent on hypoxic tolerance, to maintain cardiac function.

Neural regulation of diving responses

Interest in the mechanisms that initiate and maintain the orderly responses to breath-holding immersion had its origin in the late nineteenth century with experiments by the French physiologists Paul Bert and Charles Richet. The first recognition that cardiovascular adjustments play an essential role in the adaptations to diving is attributed to their observations of reduced heart rate in experimentally submerged ducks and chickens. Bert (1870) attributed the duck's superior underwater endurance to its enhanced oxygen reserve contained within a greater blood volume. Richet (1899) extended that observation by finding that the duck's diving endurance depended also on a reduction of its metabolic rate during dives.

The seal's reactions during dives are mediated by neural reflexes governing respiratory and cardiovascular reactions that are not unique to their species. Their expression is more exaggerated in seals, appropriate for the more demanding requirements of their diving lifestyles. The integrated responses to immersion are further subject to modification by simultaneously imposed counterstimuli. Exercise-induced faster heart rate resulting from swimming effort will likely become the dominant reaction, and dive durations will be correspondingly reduced, responding to the related more rapid changes in metabolism. However, that effect may in some demanding conditions be overridden by a persisting but more moderate bradycardia, as observed in seal swimming excursions under ice, demonstrating that the animal is functioning in combined exercising and diving response modes (Elsner et al. 1989).

Respiration and circulation, the vital functions concerned with gas exchange and distribution within the animal, are governed primarily by reflexes of the central nervous system. The mechanisms that control their functions have been the subject of study, but our present understanding of their regulation in diving mammals is assuredly incomplete. Questions remain, but the outlines of regulatory pathways have been determined and experimentally verified. The obvious responses of the diving seal include breath-holding, reduced heart rate, and decreased blood flow in most organs—with the exception of heart muscle and the central nervous system. The brain continues to be perfused, and circulation within heart muscle, the myocardium, is maintained at an intermittent and variably reduced level (Elsner et al. 1985; White et al. 1990).

The initiating trigger of the response to diving is the immersion of no more than the animal's nose in water, a reaction that is routinely demonstrable in either free diving or simulated laboratory manipulation. The combination of breath-holding and immersion precipitates the reflex sequence of dive-related cardiovascular reactions. As the dive progresses, the ensuing decline in blood oxygen and increasing carbon dioxide and acid products of metabolism further activate signals originating in neural chemoreceptor organs perfused by arterial blood (Daly, Elsner, and Angell-James 1977).

The cumulative dive effects result in the lowering of blood oxygen (hypoxemia), increasing blood carbon dioxide (hypercapnia), and accumulation of acid metabolic products (acidosis). These combined effects would normally result in an increased urge to breathe, but that stimulation is suppressed by response to the presence of water on the face of the animal (Elsner and Daly 1988). The associated neural message to brain centers acts to maintain the bradycardia and vasoconstriction despite progressive counterstimuli for the resumption of breathing (Daly 1997).

While the regulatory mechanisms that are effective in diving seals are becoming clear, those that govern respiration and circulation in other examples of "strategic retreaters" considered here are less well understood, some not at all. We can, for example, detect and test similar elements of the diving seal's regulatory reactions in breath-holding human divers, both leading to reduced metabolism, but we can little more than speculate about how the lowering of metabolism in the deeply meditating yogi comes about. A practice of repeated breath-holding episodes during meditation by some yogis, as described in chapter 3, suggests a similarity to some of the seal's diving reactions, but its relevance to depressed metabolism, while suggestive, is untested and remains speculative. The consequent decline in metabolism is consistent with mammalian responses that can be expected to arise from just this condition, but challenging variations of these basic responses remain to be examined. Progress in understanding the mechanisms that control and modulate the seal's adjustments to diving has depended upon comparing and testing how they are regulated with studies of both habitual divers and nondivers among mammals and birds.

Pronounced reduction of heart rate, bradycardia, has become a frequently cited characteristic of the breath-holding diving animal, assuming a primary place as a defining indicator of the appropriate neural reaction to submergence. It is, in reality, the readily observed portion of the more encompassing general reactions involving numerous neural circuits in an integrated response resulting in the coordination of respiratory and cardiovascular events during the dive. The abrupt sequence of events at the beginning of the seal's dive argues for its initiation by a neural mechanism rather than the slower effect that would be mediated by endocrine or chemical transport in circulating blood. The initiating trigger is the immersion of the animal's nose in water, precipitating the sequence of breath-holding and cardiovascular reactions. As the dive progresses, the ensuing decline in blood oxygen and increasing carbon dioxide and acid products of metabolism activate additional reflex signals originating in specialized nerve structures. These are the carotid bodies, small chemoreceptor organs, sensitive to the levels of blood-transported gases, located within the wall structure of the carotid arteries of the neck and perfused by arterial blood en route to regulatory centers in the brain. The cumulative effects of the lengthening dive resulting in this combination of respiratory stimuli would normally stimulate an increased urge to breathe in the nondiving animal, but that response is suppressed by the combined excitations from neural stimuli responding to the presence of water on the face of the animal combined with stimulation of neural inputs from the lungs (Angell-James, Elsner, and Daly 1981). These individual reflex responses seldom act alone, and it is the interaction of reflexes that results in the integrated diving response. The reflex actions immediately stimulated by immersion and cessation of breathing are entrained at the beginning of the dive, while slower responses to blood-borne substances (oxygen, carbon dioxide, and acid products as well as hormonal substances) take effect as the dive progresses.

The resulting impulse traffic in brainstem neurons acts to maintain by coordinated reflex mechanisms the bradycardia and peripheral vasoconstrictions typical of the diving response despite progressive counterstimuli for the resumption of breathing (Elsner and Daly 1988; Daly 1997). Similar responses have been described in diving ducks (Jones 1976; Butler 1990, 2004). Such reflex responses can also be detected in the reactions to breath-holding and immersion by nonaquatic mammals, including dogs and humans. The changes are more extensive in seal blood; for example, pH at the termination of a long duration dive may fall to 7.0, a relatively acidic value not well tolerated by a human or other terrestrial mammal. For mammals generally, divers and others, responses regulated by these neural structures protect against the hazards of accidentally inhaling water and noxious substances. The combinations of reflex actions are not simply additive; rather they interact positively or negatively depending upon the reactive condition of the animal and the particular contributions of each reflex, sometimes increasing the intensity of the response in a characteristically predictable manner.

This effect is exaggerated in seals, interacting with inputs from several sources: lung stretch receptors and the blood-gas-sensing carotid body, together involving the trigeminal, glossopharyngeal, and superior laryngeal nerves. The responses appear to result, as previously suggested, from modifications of the individual reflexes (review: Daly 1984; Elsner 1999). Similar responses to the combined reflexes take place in terrestrial mammals during breath-holding, but with a shortened time course and greater reactivity to the impending adverse consequences. These reactions and the mechanisms that control them are reviewed in publications by Daly (1984) and Elsner and Gooden (1983).

The agents that activate these responses to the decline of oxygen-based metabolism are numerous substances that operate as transmitters and modulators of the neural signals. Their sites of action are located at nerve pathway junctions, where they act to amplify or inhibit the signals that comprise the pattern of neural impulses. Their chemical composition, blood gases, and acidity vary among different connecting networks, but they are the widely distributed and frequently invoked transmitters of neural reactions. They act by modifying the electrical impulse traffic across synaptyic connections through binding to charged ions that individually enhance or inhibit transmission at the various reactive sites. What we know of the central nervous system reactions to the developing condition has depended in large part on knowledge derived from experimental studies with ducks and seals, the species known to have unusual tolerance to asphyxia such as that occurring in long dives.

The reactions governing the seal's protective responses resemble extensions of the more general mammalian mechanisms associated with varying levels of metabolic activity. A small fraction of the consumed oxygen may ordinarily not be converted to end-product elimination. This condition results in modest production of highly reactive oxidant substances, such as superoxide and hydrogen peroxide. Reperfusion of tissues that had been deprived of circulation during dives is a primary condition for oxidant production and presents a potential threat of pathological damage in the form of disrupted normal cellular metabolism (McCord 1985). Seals are endowed with enhanced antioxidant capability for dealing with this threat (Elsner et al. 1998; Zenteno-Savín, Clayton-Hernández, and Elsner 2002; Vázquez-Medina, Zenteno-Savín, and Elsner 2006, 2007). The capacity for detoxification of these potentially harmful oxidants increases with age in young seals, beginning while juveniles are maturing on land and have not yet embarked on a diving lifestyle (Vázquez-Medina et al. 2011, 2012). These results are supported by observations that juvenile seals may be seen to be perhaps preparing for initiation of the diving lifestyle by frequent periods of breath-holding before venturing to sea.

ATP, ancient energetic resource

ATP, adenosine triphosphate, is present in the tissues of all vertebrates and responds to the onset of asphyxia by its breakdown and release of adenosine. Adenosine and its precursor, ATP, are universally encountered throughout living forms from single-cell organisms to humans. The adenosine portion of the ATP molecule is made up of adenine and ribose, both components of the DNA molecule, attesting to the intimate relatedness of ATP to earth life's origins several billion years ago.

The direct application of ATP in support of cellular activity is its immediate function, but it is also the source for the synthesis and tissue storage of energy reserves in glycogen. In addition ATP functions as a signaling molecule in its role as a neurotransmitter. Its breakdown product, adenosine, is a powerful blood flow regulator, a dilator of resistance vessels. It becomes increasingly active as its concentration increases with the degradation of ATP in the process of oxidative metabolism. It plays an important beneficial and protective role in cellular viability as ischemia continues, resulting in the reduction of oxygen consumption of the blood-deprived tissues. Consequent anaerobic metabolism, not requiring a ready source of oxygen, continues, but its relatively modest contribution contrasts markedly with substrate oxidation, yielding a fraction of useful energy that would be derived from oxygen-based processes (Forman, Velasco, and Jackson 1993).

Adenosine plays a major role in responding to deficient oxygen availability and the conversion to anaerobic metabolism. The extent of the glycolytic resource varies considerably with species, being brief in humans, longer in seals, and vastly extended in turtles and other asphyxia-resistant species. The presence of adenosine is an indicator of energy deficit and of conversion to anaerobic glycolysis, its concentration increasing as ATP breakdown proceeds. Thus adenosine has a role in protecting the brain when it is deprived of blood flow and oxygen. This and other aspects of its contribution to maintaining cerebral integrity have been usefully reviewed by Lutz, Nilsson, and Prentice (2003).

The progressive change in concentrations of respiratory gases in the blood of a breath-holding animal leads inevitably toward progressive asphyxia. This combination of hypoxia, hypercapnia, and acidosis differs from that encountered at high altitude: hypoxia, hypocapnia, and alkalosis, as noted earlier. However, an untoward effect experienced at altitude, and resulting from the reduced oxygen pressure of that environment, may be suspected to occur also in the diving seal. That is the constriction of lung blood vessels in response to hypoxia that results in elevated pulmonary arterial pressure. Direct verification of its occurrence in diving animals is lacking; it has not been looked for, and they may have well developed resistance to its presence among their combined adaptations for long dives. Increased concentrations of the vasodilator nitric oxide in the lungs of human populations native to high altitudes is associated with the protective effect of their decreased hypoxic pulmonary vasoconstriction (Hoit et al. 2005). Similar or related effects may also operate in marine mammals as a constitutive component of their adaptive response to diving.

Vertebrate species vary considerably in their reactions to major disruptions of oxidative energy resources and in the quantitative expressions of their tolerances for respiratory interruption. Differences among these variations can help to clarify the regulatory mechanisms that govern their asphyxial vulnerabilities. Extreme tolerance is well illustrated in examples drawn from asphyxia-resistant fish and turtle species that can survive lengthy disruptions of their more favorable environmental conditions (review: Lutz, Nilsson, and Prentice 2003).

Multiple human and seal modulations

The study of individual reflexes, isolated from interactions with other reflexes, has provided much insight into their operations. But many of the more vital physiological reactions are governed by multiple neural reflexes, and their interactions often yield responses differing from those of the separate single reflexes. These discrete responses become more complex and unpredictable in the living animal by their interactions in combination with other concurrently active reflexes. The neural mechanisms governing specific diving reactions are discrete reflex responses, but it is well to recognize that such actions seldom, if ever, occur as isolated reactions independent of other effects that may alter the animal's responses. The seal's underwater submersion is frequently accompanied by swimming exercise, as in pursuit of prey, interaction with other animals, and escape from perceived threats. The associated reflex responses that govern these effects are subject to modification and interaction with other coincident influences, both external to and internally activating other and unrelated reflex reactions. Our understanding of mammalian respiratory and circulatory neural regulatory responses has made progress by examination of discrete reflexes in isolation from other influences likely to interfere with their effects. Such is the nature of research techniques that we clearly make better progress toward resolution of individual topics of study by isolating them from potentially competing or disturbing influences. Nature may often operate differently, however; isolated and single reflex actions are relatively rare, while interactions among reflexes and other modifying influences are frequent and routine.

The customary and productive experimental approach to understanding how living systems work has been to reduce the components of each system to its simplest separate parts. That is the approach of reductionism, rendering complex biological reactions into components that are manageable for discrete experimental manipulation. This technique has been the primary and productive method of physiological investigations. It works well in achieving discrete analysis of individual reflexes and their effects, but it has been less effective in revealing how the parts interact and combine to form the complex reactions of living systems. The regulatory mechanisms described here are subject to modification by a variety of competing influences depending on species, physiological state, and environmental conditions. Human dive reactions, as noted earlier, resemble in some respects the cardiovascular responses of seals, despite their brief and much more modest expression. Their slower onset, requiring several seconds to develop in contrast with their often abrupt initiation in seals, suggests that the human response is more dependent upon the gradual change in blood gases and pH toward those of the asphyxial condition.

The separate responses of humans to breath-holding alone and to immersion in water differ in magnitude, breath-holding without immersion resulting in lessened responses in most subjects (Elsner and Gooden 1983). Combined effects of breath-holding and face immersion result in an intensified reaction. In a more general application than that restricted to diving, these reflex responses are related to those that protect against accidental inhalation of water, smoke and other disturbing or noxious substances (Daly 1984; Elsner and Daly 1988). Reflexes are subject to numerous interactions that alter or suppress the effects that may be predicted from actions restricted to their effects in isolation. Some reactions of diving animals and their often unpredictable interactions with other reflex inputs have been identified and described by Daly and coworkers (Daly 1984). We excel in resolution of isolated experiments, sometimes ignoring that the isolation is of our doing, not the way the natural and undissected system works. The dilemma for the investigator is well stated by Alvin Toffler (1970): "One of the most highly developed skills in contemporary Western civilization is dissection: the split-up of problems into their smallest possible components. We are good at it. So good, we often forget to put the pieces back together again."

These effects have been described by Daly (1997), referring to the conditions operating in the intact animal: "What are also important in our understanding of the overall control of the cardiovascular system are the ways two or more reflex systems interact together, for in any disturbance of the circulation hardly ever is only one group of receptors affected on its own. In this respect our knowledge is still meager. The aim should be to analyze each reflex quantitatively and to superimpose another input in a controlled way, at the same time maintaining constant, or as near constant as possible, other physiological variables."

The autonomic division of the nervous system is intimately involved with and plays an essential role in the initiation and interactions of the diving responses. Its role in activation and modification of visceral reactions such as cardiovascular effects, temperature regulation, and respiration has traditionally been assumed to be subordinate to that of the integrative role of controls based in higher brain centers. Specifically, the learning of autonomic effects was regarded as of little consequence and received little attention. That view was much modified and dispelled by the work of Neal Miller and colleagues (review: Miller 1969). Their research attention was concentrated on the accessibility of the autonomic nervous system to the effects of learning.

Some of the regulatory mechanisms that operate in diving seals have been revealed; many of the other examples of depressed metabolism are less well understood, some not at all. We can, for example, detect and test similar but much less responsive elements of the diving seal's regulatory reactions in breath-holding human divers, but we lack knowledge of how meditating yogis lower metabolism or what governs the precise timing of hibernation intervals. A promising avenue of investigation was suggested by evidence that activation of autonomic nervous system responses is subject to manipulation by learning (Miller 1969). A persisting problem with these studies is that the topics of interest, cardiovascular responses, for example, are also subject to modification by subjective or emotional activation by inadvertent stimuli. It may, therefore, be difficult to distinguish the autonomic reactions from incidental or voluntary influences altering the responses under study. Miller's analytical procedures have done much to clarify and circumvent this potential problem.

These and other experimental studies have shown that the mammalian autonomic nervous system is readily trainable, and they raise inevitable provocative questions regarding the specificity of the circulatory and metabolic responses to diving. Little research has been done on this topic; one exception is a study by Ridgway, Carter, and Clark (1975) demonstrating that a nondiving California sea lion could be trained by operant conditioning to react on command by slowing its heart rate to ten beats per minute. It is, of course, clear that the cardiovascular responses are essential for the animal's ability to survive long dives, but the variability and trainable nature of these autonomic reactions suggest the need for further study of their modifications by accompanying influences other than diving.

The preconditioned seal?

Preconditioned seems especially appropriate for a description of the frequently diving seal, as will be suggested in the next chapter. We cannot, of course, assume that seals occupy a special place in the continuum of mammalian adaptations in this regard, however much they may appear to be prime candidates for implementation of its favorable characteristics for the natural diver. Its operation in seals has not been tested. Preconditioning, or something closely resembling it, appears to be the likely reaction of the diving animal to its frequent repetition of submergence and the consequent suspension of respiration and variations in blood flow distribution.

THE

CONDITIONING

PHENOMENON

The dynamic interactions of mammalian respiration and the circulation of blood are inextricably linked in a closely functioning association. Their smooth coordination responds with increase or decrease in reaction to the varying requirements for regulated support of regional metabolic activity. Marked reduction of the circulation similar to that initiated by the diving seal, as described in chapter 2, may be found to have another effect in addition to its characteristic response to lessened metabolic demands.

The consequence of repeated brief interruptions of regional circulation resembling those that are characteristic of circulatory responses to diving has been identified as the phenomenon of cardiovascular preconditioning (Murry et al. 1990). This term has been applied to the unexpectedly improved cellular survivability and reduced metabolism that result from the induction of brief periods of decreased or arrested blood flow. Tolerance of a subsequent circulatory deprivation is much improved by this simple procedure. The reaction is demonstrable in two phases, an immediate one lasting a few hours and a subsequent effect delayed for hours to days. Research for more than two decades has drawn attention to improved viability of diverse organs resulting from this counterintuitive process. It has been produced in virtually every mammalian species tested, but the mechanism of its induction remains not well understood.

Preconditioning may be induced in most organs, including heart, liver, muscle, brain, and intestine. The initial tissue exposures to lack of circulation are applied sequentially, lasting a few minutes each, during which blood flow is prevented or much reduced, alternating with restoration of normal perfusion. Cellular reactions are so modified by these brief cycles of circulation, resulting in the inexorable alteration of downstream tissues toward lower oxygen, increased carbon dioxide, and increasing acidity, as to render them tolerant of subsequent more prolonged blood flow reduction.

There are experimental observations in many animal species that suggest conditioning to be a widespread, perhaps universal, phenomenon. Specific tests of conditioning have not been extended to marine mammals, but they may have special relevance to these species due to their characteristically reduced circulatory and metabolic reactions during dives. The accompanying frequently repeated cycles of blood flow alteration suggest that naturally occurring conditioning may be a regular feature of the diving seal's lifestyle.

Conditioning effects

Reduced metabolism is a general consequence of the conditioning reaction. An example is the decline in energy demand of the preconditioned heart indicated by decreases of both aerobic and anaerobic metabolism demonstrated by reduced lactate production and delayed ATP metabolism. Preconditioning was found to reduce the metabolism of the dog heart by 40 percent in some circumstances (Murry et al. 1990). Accordingly, metabolic activity supported by consumption of glycogen, the tissue storage of carbohydrate, is attenuated in the preconditioned organ (figure 6.1).

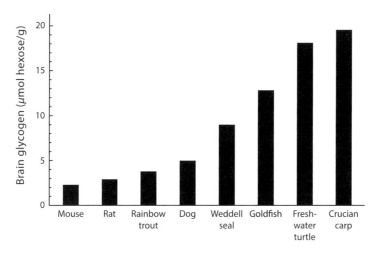

6.1.   Brain glycogen contents in several vertebrate species. Redrawn from Lutz, Nilsson, and Prentice 2003; Kerem, Hammond, and Elsner 1973.

The primary effects of preconditioning are demonstrable during a brief period limited to no more than a few hours after their initiation. The delayed phenomenon, a longer-lasting second window of activation, appears some twelve to seventy-two hours after the original preconditioning effect (Kuzuya et al. 1993). Its protective benefits are somewhat less effective, but they persist for several days. Immediate and delayed effects of enhanced ischemic tolerance are repeatable with application of the original circulatory interruption. Both neural and endocrine regulatory processes have been identified, but the mechanisms involved remain incompletely understood.

Consideration here of the conditioning effects described is based on their possible contribution to the metabolic depressions that are characteristic of both diving seals and meditating humans. Diving seals are suggested to be likely prospects for preconditioning reactions, although that phenomenon has not been looked for in them. Their frequent diving and associated reductions in both total and regional circulation appear to make them appropriate candidates for conditioning effects. The physiological responses of the meditator are less certain prospects. However, some possibilities for such effects exist in the repeated deep depressions of metabolism such as those shown to occur in dedicated practitioners.

Another conditioning effect appeared with the later discovery of some beneficial effects of *postconditioning* (Zhao et al. 2003), briefly repeated episodes of tissue ischemia alternating with perfusion initiated during recovery from an extended period of tissue ischemia. This response results in increased resistance to an extended episode of interrupted blood flow followed by intermittent restoration. The procedure may hold more favorable promise of clinical applicability than preconditioning, because of its potential use during restoration of blood flow after circulatory deprivation as, for example, during resumption of cardiovascular function following invasive procedures that require temporary circulatory arrest. The simplicity and profound effects of the conditioning phenomenon have influenced attitudes and thinking with regard to cardiovascular physiology and pathology. These disarmingly simple responses have led to important new insights into the nature of some apparently fundamental and previously unexplored circulatory responses.

Interest generated by the quest for the mechanism of cardiovascular conditioning reactions naturally arouses some questions relating to seal biology. Seals, by virtue of their circulatory alterations during diving, engage in frequent and lengthy bouts of selective vasoconstriction and reperfusion and consequent cyclic ischemia and restoration of blood flow widespread in organs and tissues. The natural history and implications of pre- and postconditioning can be appreciated in context when compared with what comes naturally for diving mammals and the possibility that preconditioning may play a role in the expression of their diving facility. The seal's circulatory adaptations for diving clearly resemble the reactions conducive to preconditioning, but no direct involvement of preconditioning has been identified (nor looked for) in the seal's adaptive repertoire.

Conditioning mechanisms

Preconditioning was first identified as a consequence of intermittent myocardial perfusion in dogs (Murry, Jennings, and Reimer 1986). Complete cessation of circulation is apparently not essential if the imposed blood flow reduction is sufficient (Shizukuda et al. 1992). The effect can also be produced remotely; repeated brief applications of a limb tourniquet can induce cardiac preconditioning, for example. Repeated brief interruptions of blood flow result in little cumulative ATP degradation and less consequent acidosis (Reimer, Vander Heide, and Jennings 1994). Confirmations of reduced metabolism have been identified by Jennings et al. (2001). Their summarizing statement suggests the potential for useful clinical application: "The results support the hypothesis that a reduction in energy demand is an essential component of the mechanism of cardioprotection in preconditioned myocardium. However, the mechanism through which ischemic preconditioning results in lower energy demand remains to be established."

An episode of brief preconditioning renders the dog's heart considerably more resistant to subsequently reduced blood flow, an intervention sufficient to produce cardiac cell damage before the conditioning episode. The preconditioning sequence so improved the tolerance of the heart that an experimentally induced circulatory arrest could be increased in extent by 75 percent, protection far exceeding that of currently employed therapeutic interventions (Yellon and Downey 2003). If a similar conditioning effect occurs in the seal's heart and other organs while experiencing circulatory restriction and reperfusion during dives, still untested, it may be expected to represent a considerable metabolic savings of likely benefit to the submerged animal.

Preconditioning effects vary with species, some requiring longer or more frequent exposures (Kloner and Jennings 2001; Tsang, Hausenloy, and Yellon 2005). Some tests have been successful with a single ischemic interval no longer than a minute. Cardiovascular conditioning may have an ancient evolutionary history, demonstrable in fish as well as mammalian species. Similar reactions to brief hypoxic exposures have been noted in other experiments, including the temporarily reduced metabolism of goldfish (Prosser et al. 1957) and trout (Gamperl et al. 2001). These effects in both mammalian and non-mammalian species support the notion that preconditioning is a general vertebrate reaction.

Transient ischemia in adjacent portions of the heart and in remote noncardiac tissue is reported to reduce the pathological impact of interrupted local blood flow in cardiac muscle. That is, subjecting the kidney, intestine, and other organs to brief cessation or reduction of circulation protects them and other remote organs from ischemic damage during subsequent more intense and longer exposure (Pell et al. 1998; Bonventre 2002; Cohen, Baines, and Downey 2000; Kerendi et al. 2005). The preconditioning effects resemble in these respects the favorable responses resulting from hypoxic exposure noted in earlier studies of the mammalian heart (Poupa et al. 1966; McGrath and Bullard 1968; Meerson, Gomzakov, and Shimkovich 1973). Broader implications of its applicability refer to other organ systems as well.

Several substances appear to be involved in a redundant combination of mediators. One of the more likely of these is adenosine, the product resulting from metabolism of adenosine triphosphate (ATP), the universal cellular compound that supplies high energy for vital reactions. The evidence suggests involvement of adenosine receptors and neural reactions in its initiation (Gho et al. 1996; Granfeldt, Lefer, and Vinten-Johansen 2009). Among the numerous metabolic products present, adenosine may occupy a premier position as both breakdown product of ATP metabolism during ischemia and a depressor of metabolic activity (Jennings et al. 2001). Adenosine has been found to reduce cellular oxygen consumption in a wide variety of species from primitive to mammalian (review: Lutz, Nilsson, and Prentice 2003; Karimi, Ball, and Power 1996). By this effect, it may be especially important in providing protection when required by the hypoxia–challenged heart. Adenosine is so rapidly metabolized that its potential metabolic reactions may be limited, but some agents having similar properties have been shown to have related therapeutic value (Cohen and Downey 2008).

Additional metabolic effects attended the discovery of remote preconditioning, a regional protective effect resulting from brief blood flow restriction in a distant organ or a separate portion of the same organ (Kerendi et al. 2005). This prospect argues for elaboration of a blood–borne substance that acts through transport in the general circulation. Considerable evidence supports the view that adenosine is prominently involved, acting directly or through an intermediary mechanism (Liem et al. 2002; Yellon and Downey 2003; Hausenloy and Yellon 2008b). The regular appearance of lactate, resulting from anaerobic metabolic activity even early in a dive (Kjekshus et al. 1982), suggests that similar conditions may exist regularly in seals. Increased levels of hypoxanthine, the adenosine breakdown product, in ischemic

seal heart and kidney suggest indirect experimental evidence of related lactate production in these conditions (Elsner et al. 1998).

Hypoxia and conditioning

Recent research into the conditioning phenomenon and reconsid-
erations of earlier experimental results have revealed that what has
been regarded as "chronic" hypoxia has in some instances actually
been "intermittent" hypoxia, due to the techniques employed in the
housing and care of the experimental animals used. Apparently, the
difference may be important, intermittent exposure and the related
frequent reoxygenation being more effective in producing hypoxic
preconditioning (Cai et al. 2003; Milano et al. 2002, 2004). Experimental modification of cardiac function by such procedures as intermittent hypoxia or brief reduction of coronary blood flow suggests
a possibility for beneficial results. Related studies in animals and humans have been cited by Zong et al. (2004), noting the apparent similarities in cardiac protection against infarction resulting from either
intermittent hypoxia or ischemic preconditioning.

The mechanisms whereby pre- and postconditioning derive their
protective effects have not been resolved; new examples and interpretations continue to be revealed. One of the more compelling of these
is the confirming demonstrations that the viability and reactivity of
organs and tissues can be supported and enhanced by such simple manipulations of their circulations. Discovery of the conditioning phenomenon has attracted much research attention since its original identification by Murry, Jennings, and Reimer (1986), with special concern
for its potential clinical applicability to treatment of heart disease.

The preconditioning paradigm has brought considerable attention to the way in which restricted circulatory perfusion of diverse
organs results in conferring resistance to subsequent ischemia. This
altered resistance produced by the simple manipulation of briefly and
intermittently restricting blood flow resembles in some respects the
reaction produced by intermittent exposure to hypoxia. Prospects
for its beneficial effects have been directed specifically to the heart
(Poupa et al. 1966; Poupa 1976; McGrath and Bullard 1968; Meerson
et al. 1973). Zong et al. (2004) tested its effect on cardiac function in

dogs by exposing them for brief intermittent periods of ten minutes or less to approximately one-half sea-level oxygen pressure during a twenty-day period. A subsequent sixty minutes of coronary occlusion was ineffective in producing myocardial infarction, protection that was lacking in untreated control animals. The improvement was not attributable to hypoxic acclimatization, because the exposures were not sufficient to result in increased circulating hemoglobin or other changes characteristic of chronic hypoxia.

Exposures to intermittent hypoxia clearly differ from experimental treatments of brief organ ischemia in preconditioning, but the similarity of the resistances produced by repeated episodes of hypoxia on cellular tolerance raise obvious questions regarding possible common or related protective mechanisms. Such beneficial hypoxic exposures have received less research attention than those resulting from preconditioning, and prospects for revealing whatever common features might exist suggest an attractive topic for further investigation.

Biomedical implications

Inquiry into the nature of pre- and postconditioning effects, potentially beneficial and more easily applied than other cardiovascular interventions, have inspired a vigorous pursuit of practical medical applications for the technique, quite aside from what their elaboration might tell us about fundamental physiological processes. These efforts have thus far been rewarded with fewer than expected realistic clinical opportunities, and prospects for success continue to be examined. Current cardiology research endeavor has devoted much attention since its discovery to an attempt to gain insight into the protection of the heart that is produced by the mechanism of ischemic or hypoxic conditioning. Possible clinical relevance of the conditioning reaction has attracted considerable attention (Hausenloy and Yellon 2008a; Granfeldt, Lefer, and Vinten-Johansen 2009). Reviews of some clinical applications indicates successful use of remote ischemic preconditioning of a different organ (Kloner 2009; Granfeldt, Lefer, and Vinten-Johansen 2009).

Much attention has been devoted to the potential clinical implications, especially those pertaining to the heart, of the conditioning phenomenon (reviews: Kloner and Jennings 2001; Tsang, Hausenloy, and Yellon 2005). Implications of cardiac conditioning have introduced a new sense of hope for powerful tools with which to confront this leading cause of premature death in the Western world. The prospect of applying preconditioning to altering the course of human heart disease has been the stimulus for considerable attention and investigation devoted to understanding the nature of the conditioning response. Several thousand related research publications have appeared since preconditioning was first described by Murry, Jennings, and Reimer (1986).

The primary drive that has fueled these studies has been the prospect that an important novel attack on the scourge of heart disease may be at hand. The potential clinical importance of cardiovascular conditioning derives from the observation of a marked improvement in organs' resistance to lack of oxygen induced by brief deprivations of circulation. These positive reactions suggest the promise of a new paradigm for treatment of cardiac (Bolli 2006) and other circulatory conditions (Granfeldt, Lefer, and Vinten-Johansen 2009). Some benefits to be derived from its clinical application have been realized, but they have still to result in the prospects that have been anticipated for treatment and preventive medicine (Ludman, Yellon, and Hausenloy, 2010).

Preconditioning has been deemed inappropriate for some therapeutic applications, because of the necessity for prediction of pathological cardiac events during which its ameliorating effects could be realized. But while the preconditioning reaction has opened new windows into cardiovascular reactions, general clinical application has been frustrated by the need to apply the technique in anticipation of an ischemic event and the unrealistic nature of such prediction. Suggestions of its dependence on an undiscovered circulating substance or connection have failed to reveal its origin despite vigorous research efforts. Despite the frustration related to these deficiencies, much new insight has been gained concerning unanticipated aspects of blood supply to heart muscle and more generally to numerous other organs. Interest in the concept is indicated by the numerous related publications that have appeared since its original 1986 description. Intense research efforts looking into its mechanism continue to make progress.

Better prospects appear to result from more recent developments that identify equally effective myocardial protection from postconditioning, brief intermittent blood flow during reperfusion after a period of occlusion (Zhao et al. 2003; Sun et al. 2005; Gerczuk and Kloner 2012). Positive cardiac effects resulting from conditioning in peripheral parts of the circulation, intermittent restriction of limb blood flow, for example, have also been identified. This aspect of the postconditioning phenomenon has received special attention because of its practicality and anticipated applicability for patients of heart disease. Useful application of postconditioning has quite naturally centered on its possible use in the management of cardiac conditions, but its involvement in other parts of the circulation also appear to warrant more general examination. Postconditioning has seen limited clinical application (Zhao et al. 2003; Tsang, Hausenloy, and Yellon 2005; Hausenloy and Yellon 2007; Andreka et al. 2007).

Implications of the conditioning process for understanding some basic physiological attributes and for medical practice of cardiology are discussed in a recent publication (Gerczuk and Kloner 2012), a review of therapeutic prospects for limiting the severity of myocardial infarction. Remote ischemic postconditioning can be as simple as inducing controlled reductions of limb blood flow. Such a noninvasive application of therapeutic intervention after an ischemic cardiac event has been shown to have a clearly positive effect equal to or superior to that of other remedial applications. This simple technique has been successfully applied in emergency situations and during transport to treatment centers. Application of tourniquet–reduced limb blood flow as a simple first-aid measure, referred to as *perconditioning*, has been successfully employed in ambulances responding to the aid of heart attack victims (Schmidt et al. 2007).

Reviving the failing heart

The literature of cardiology has for many years elaborated on the paradoxical hazards of reperfusion in human heart muscle despite the apparent benefits of blood flow restoration after coronary infarction and the renewal of its circulation. Nutritive resupply and washout of accumulated noxious metabolic products, both actions vital to survival of cardiac muscle cells, also pose the threat of new insult originating from excess production of potentially hazardous free radicals generated by the reintroduced flood of excessive blood-bound oxygen consequent to the restoration of perfusion. They pose a threat of potential cell damage in what is referred to as *cardiac stunning*. Some beneficial effects of briefly maintaining cardiac tissue acidosis during recovery from hypoxia have been recognized for several decades (Bing, Brooks, and Messer 1973). This improvement is dependent upon the presence of acidic tissue pH during reperfusion (Kitakaze, Weisfeldt, and Marban 1988; Cohen, Yang, and Downey 2007), as occurs in the progressive return to normal pH during restoration of blood flow.

Protective cardiac reactions to hypoxia resembling in some respects those of preconditioning had been observed for several years before the specific identification of effects by Murry, Jennings, and Reimer (1986). These beneficial responses were suggested by Poupa et al. (1966), McGrath and Bullard (1968), and Meerson, Gomzakov, and Shimkovich (1973). Exposure to intermittent hypoxia confers cardiac protection from a subsequent potentially detrimental hypoxia exposure in rats (Béguin et al. 2005) and dogs (Zong et al. 2004). The beneficial effects result from the generation of an acidic tissue condition during reperfusion (Kitakaze et al. 1997; Zong et al. 2004; Cohen, Yang, and Downey 2007). These observations, indicating that novel and beneficial treatments might be derived from this source, have created renewed interest in the continuing research efforts. They suggest that a circulating agent or a neural pathway exists through which the preconditioning reactions are mediated, opening the prospect that it may be possible to more fully exploit this approach for clinical use.

The prompt negative reactions of air-breathing organisms to oxygen lack are abrupt and severe. Modest restriction of oxygen intake may, however, be well tolerated, and it can result, counterintuitively, in advantageous benefit in the form of enhanced cardiac function (Ostadal and Kolar 2007). The reaction of exposure to intermittent hypoxia may render the heart more tolerant of ischemia than hearts lacking such treatment. Intermittently breathing low oxygen has been noted to confer more protection than chronic hypoxia alone (Asemu et al. 1999; Milano et al. 2002, 2004). Somewhat similar cardiac benefits of exposure to high–altitude hypoxia have been recorded (Heath and Williams 1981; Neubauer 2001).

Considerable uncertainty exists with regard to the mechanism and origin of the protections conferred upon cardiac tissue by intermittent hypoxia or by pre- and postconditioning. Still, the idea persists from these studies that possible beneficial effects from such treatment lead to improved cellular viability. Repeated brief exposures of dogs to hypoxia, insufficient to result in hypoxic acclimatization, led to increased collateral blood flow in heart muscle. Some progress has been made in the suggested promise of related useful clinical applications (Granfeldt, Lefer, and Vinten-Johamsen 2009; editorial: Kloner 2009).

Preconditioned seals and yogis?

The diving seal's cardiovascular responses (drastically reduced heart rate and multiple organ vasoconstrictions) suggest that they are appropriate circumstances for initiation of conditioning responses. Modest reductions in body temperature, similar in magnitude to those frequently occurring in seal dives (Scholander, Irving, and Grinnell, 1942b; Hammel et al. 1977), enhance the protective effect (Chien et al. 1994; Gho et al. 1996). As a consequence of these lines of evidence, the primary reactions of diving seals are likely to make them good, perhaps ideal, candidates for these effects and for conditioning as a likely support for their diving activities.

Seals and other diving species have not been specifically tested for the possibility that their recovery from regional ischemia after dives results in cardiovascular preconditioning. The conditioning effect following upon a period of tissue blood flow reduction is pertinent to the situation of the diving mammal after extended periods of breath-holding dives. Resumption of blood flow results in a brief period of maintained tissue acidity associated with the presence and accumulation of metabolic products. The acidic condition of the reperfused tissue is generally accompanied by some production of oxygen radicals, the highly reactive forms of oxygen. Their brief appearance during reperfusion has been found to contribute to protection by preconditioning (Cohen, Yang, and Downey 2007).

The measured pace at which circulatory perfusion is resumed during postconditioning results in maintaining an acidic blood and tissue condition while blood flow is being gradually renewed. This maintained acidic pH is an apparent requirement for revival of the ischemic cells, a condition referred to as the *pH paradox*, so named because of its apparent contradiction with the restoration of normal circulation (Currin et al. 1991). The course of circulatory events in postdiving seals resembles rather strikingly this protective role as described in the postconditioned animal and human heart (Cohen, Yang, and Downey 2007; Hausenloy and Yellon 2007; Vinten-Johansen et al. 2007; Granfeldt, Lefer, and Vinten-Johansen 2009).

Deprivation or reduction of kidney and other organ blood flow, sometimes for many minutes, and subsequent reperfusion are well tolerated by seals, and this condition persists with momentary restorations of flow throughout sequential dives (Elsner et al. 1966). Such frequently repeated episodes demonstrate a level of tolerance of circulatory suspension that is unusual in nondiving animals. Study of isolated blood-perfused dog and seal kidneys showed that they differ markedly during recovery from circulatory deprivation. Seal kidneys maintained urine production, while similar treatment of dog kidneys resulted in functional failure (Halasz et al. 1974).

Consideration here of the conditioning effects described is based on speculation regarding a possible contribution to the metabolic depressions that are characteristic of both diving seals and meditating humans. The diving seal is a likely prospect for preconditioning reactions; its frequent diving and associated reductions in both total and regional circulation appear to be appropriate signals for conditioning effects. The physiological responses of meditators suggest that they are less certain candidates. The repeated deep depressions of metabolism such as those shown to occur in the examples of dedicated practitioners may, however, have a counterpart in similarly altered regional circulation.

Relevance of the preconditioning phenomenon to reactions during meditation is untested, but tenuous relationships are suggested. It is noteworthy that there exists among some dedicated yogis the conviction that the meditative condition is enhanced by simultaneous periods of prolonged breath-holding. As mentioned in chapter 3, a frequently invoked component of the meditating condition reported in some Indian yogis is the inward repositioning of the tongue with the intention of providing a blockage to the respiratory airway (Anand, Chhina, and Singh 1961). This practice of repeatedly interrupted breathing during meditation suggests that the depressed metabolism of highly experienced yogis might be partly attributable to the frequently repeated occurrence of breath-holding episodes during meditation.

Oxidative stress

Oxygen consumption by mammals and birds, essential for life-supporting energetic processes, is accompanied by the unavoidable production of highly reactive substances collectively known as *oxidants*. The quantity of this respiratory byproduct is usually small and innocuous, about 0.1 percent of the overall oxygen consumption. This small fraction of the respired gas is ordinarily diverted to the formation of highly reactive oxygen forms, such as superoxide and its enzymatic conversion by superoxide dismutase to hydrogen peroxide. Their production can be induced by abrupt reduction of oxygen available for cellular metabolism (Guzy and Schumacker 2006). Consequently, the vital process of oxidation proceeds optimally within the constraints of a relatively narrow range of oxygen concentrations (Semenza 2007). Restoration of blood flow after a period of ischemia, as occurs during the immediate recovery after a dive, is one of the conditions that may result in increased oxidant production that exceeds cellular antioxidant capacity, the condition of *oxidative stress*.

Marine mammals appear to tolerate these conditions that might be expected to lead to adverse effects in terrestrial mammals (Elsner et al. 1998; Hermes-Lima and Zenteno-Savín 2002; Zenteno-Savín, Clayton-Hernández, and Elsner 2002; Vázquez-Medina, Zenteno-Savín, and Elsner 2006). Some evidence indicates that diving mammals may have elevated antioxidant defenses against oxidative stress. In a study comparing the effects of ischemia in seal and pig tissues, production of hypoxanthine, a product of ATP degradation, was higher in pig heart and kidney than in seal (Elsner et al. 1998), suggesting that the seal may function with an enhanced capacity for dealing with such potentially adverse oxidant effects.

Superoxide dismutase and other oxygen-radical scavenging activators are elevated in seal tissues, indicating a likely mechanism for enhanced protection against reactive forms of oxygen (Wilhem Filho et al. 2002; Vázquez-Medina et al. 2005, 2007). Reperfusion of tissues that have been deprived of circulation for an extended period may be expected to result in increased local production of reactive oxygen species that would be damaging to tissues previously deprived of circulation. However, the simple technique of postconditioning, by modifying the effects of intermittent brief recirculation episodes, is likely to reduce or eliminate that hazard (Sun et al. 2005).

The seal's defenses against suspected frequent exposures to augmented levels of potentially harmful oxidants have been examined in several studies. While these products are potentially damaging, in moderate concentrations they have been found to be beneficial as an effective trigger of preconditioning detected in the rabbit's heart (Baines, Goto, and Downey 1997). Whatever protective mechanisms may be operating are not well understood, but increased endogenous antioxidants are apparently important contributors to the prevention of potential damaging effects. Notably, the naturally occurring antioxidant glutathione is highly active in seal tissues (Vázquez-Medina, Zenteno-Savín, and Elsner 2007), and it is a likely contributor to the seal's protection against oxidative damage (Vázquez-Medina et al. 2011, 2012; Zenteno-Savín et al. 2012).

The mechanisms upon which cardiovascular conditioning depends continue to be only partly resolved; many questions remain. Consideration of what appears to be the remarkable innovation of the conditioning reactions reveals new ways of looking at adaptations of which the mammalian organism is capable, while at the same time bearing intriguing prospects for clinical benefit. Among the numerous reports of experimental results suggesting prospects for useful clinical applications, some relate to possible effects that may function to invoke the responses observed in marine mammals. The conditioning reaction has not been looked for in seals, but they may well depend upon its effects for diving facility and survival.

HIBERNATION

AND DIVING

The discourse regarding animal and human temporary withdrawals from regular levels of metabolic activity has centered upon the examples of the seal's quiet nonexercising dives and humans in deep meditation. The evidence shows that these conditions may be expected to impose reductions in the rates at which the metabolic reactions in the cited examples take place. Effects that accompany animal hibernation relate to the more lengthy condition in which metabolic depression similarly plays the essential role.

Metabolic flexibility, as shown by the ease with which some animals respond to environmental conditions that threaten their livelihood, is well demonstrated in those mammals that retreat into seasonal hibernation. Recourse to this state of suppressed activity is practiced by a variety of species in several taxonomic groups. Their survival depends on their ability to avoid the metabolic costs of struggling to continue feeding and maintaining thermal stability by withdrawal into a state of metabolic depression. Restorative sleep represents another withdrawn condition among mammals, but one of much more modest metabolic decline.

The reactions of hibernating mammals to environmental impacts clearly differ in several respects from those of diving seals and human meditators, primarily in the time course and extent of their metabolic reductions and their tolerance of lowered body temperatures. Reactions to these two conditions, diving and hibernation, contrast markedly in rates of onset and duration. The resulting effects differ in the hibernating mammal and diving seal, one to initiate retreat from active life when seasonal conditions threaten food resource, the other to protect the animal's endurance of breath-holding dives.

The primary defining characteristics of both diving and hibernation rest on their profound reductions in metabolism, while their rates of onset and recovery are clearly much different. Most mammalian dives last a few minutes, though they can exceed an hour in some species, and the onset and end of their protective reactions usually occur abruptly at the beginning and termination of the dive. In contrast, hibernation in one form or another lasts hours, days, weeks, or months, varying among conditions and species, all involving a lowering of body temperature. This form of physiological depression induces metabolic conservation essential for survival by many mammals and birds in response to such conditions as food scarcity, dehydration, and seasonal climate change.

An aspect of the hibernator's and the diver's reactions identifies their dependence upon a metabolic similarity. Heller (1988) describes it in these words: "There is one physiological characteristic that these two hypometabolic states have in common: they both involve an acidosis. In both states acidosis is likley to have severely inhibitory effects at the cellular level. It may well be that an important primary regulatory event in both diving and torpor hypometabolism is a shift in central chemosensitivity."

Metabolic depression

The recourse to lowered metabolic rate of the mammalian diver raises questions regarding possible comparisons with other modes of metabolic depression. Three natural dormant conditions in animals of several taxonomic identities have been identified as sleep, shallow torpor, and hibernation (Heller et al. 1978).

Sleep consists of the essential restorative rest, about eight of each twenty-four hours in adult humans and of variable duration in other mammals. Shallow torpor occurs in small mammals and birds, usually during sleep with lowering of body temperature a few degrees, sometimes more in hummingbirds and hamsters. It is generally nocturnal, an exception being the daytime sleep of bats. Hibernation refers to deep seasonal torpor, generally maintained during several days repeatedly or of longer continuous duration. Most hibernators are small, and their body temperatures during hibernation may approach freezing. Bears, in contrast, may hibernate for several winter months, but with only a modest body temperature depression.

Mammalian tolerance of sometimes extreme but benign body temperature reduction without pathological disruption of normal life-sustaining functions is represented by the natural hibernators. In contrast, moderate cooling of more than a few degrees in nonhibernating mammals can result in cardiac fibrillation, although it may be prevented or lessened by appropriate anesthesia. It has also been shown to be prevented if the blood pH is allowed to conform normally to the increasing alkalinity that occurs with decreasing temperature rather than artificially maintaining pH at its normal unmodified value (Rahn, Reeves, and Howell 1975).

Some bird species experience several hours of night torpor during which metabolism is suppressed and body temperature declines a few degrees, but avian hibernation is reportedly limited to one species, the poorwill. Estivation represents metabolic withdrawal from dehydration during summer practiced by vulnerable small mammals, snails, toads, and frogs; diapause is a similar condition in insects.

Metabolic challenges

A general response of most animals to an encounter with low atmospheric oxygen is a reduction in metabolism. It is especially noticeable in lower vertebrates and newborn mammals. Study of the phenomenon has a long history stemming from the original observations of

Julien Jean César Legallois and Paul Bert in the nineteenth century. Small mammals are especially subject to decline in metabolism, up to 20 percent, when rendered hypoxic. Restoration of their normal oxygen consumption may not result in a compensatory increase in accumulated anaerobic "debt" (Fahey and Lister 1989; Frappell, Saiki, and Mortola 1991). The effect is likely related to their higher rates of oxygen consumption in normal resting conditions (Mortola 1993, 2004). It has been suggested that the lesser effect in large adult animals results from their more modest response expressed as a fraction of total metabolism (Mortola 2004).

The mammalian brain is highly vulnerable to the damaging effects of a decrease in its oxygen supply, as may occur in accidental restriction of cerebral circulation or an imposed reduction of respiratory function. It is, however, capable of adaptive change sufficient to improve its tolerance of exposure to low oxygen. Bickler (2004) describes two categories of such modification affecting mammalian survivability in hypoxic exposure. These are identified as acclimatization to high altitude and the hypoxic tolerance of fetal and newborn mammals.

Maintenance of a steady high metabolism and its associated warm body temperature is a challenge for animals during periods of food scarcity or reduced foraging prospects, as in winter and in desert conditions of severe water shortage. Hibernation represents a coordinated effort to conserve energy by regulated metabolic depression and the consequent reduced need for food and water intake. It is a recourse to avoiding direct confrontation with cold exposure by escape into metabolic depression in insulated nests in which the animal's slowed metabolism can be maintained at minimum expense. These and other aspects of the hibernating condition are the subjects of comprehensive reviews (Carey, Andrews, and Martin 2003; Geiser 2004).

Hibernators

Hibernators exist in several orders of mammals; nearly all with the exception of bears and marmots are of small body size. They include some bats and marsupials. A common feature is one of coordinated retreat from unfavorable seasonal climate and reduced access to or

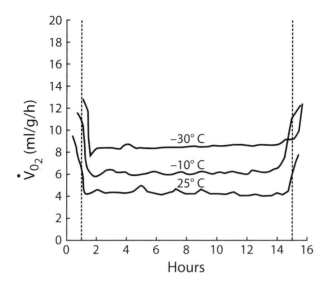

7.1.   Overnight oxygen consumption of a seasonally acclimatized black-capped chicka-dee (*Parus atricapillus*) at three temperatures. Redrawn from Sharbaugh 2001.

availability of food sources (reviews: Kilduff et al. 1993; Carey, An-drews, and Martin 2003; Geiser 2004).

Heat loss occurs more rapidly in small than in large animals, be-cause they have a greater ratio of surface area to volume, higher spe-cific metabolic rate, and less body insulation. This thermal charac-teristic influences the hibernating character of bears and to a lesser extent marmots, resulting in an exception to the general rule relating small animal size with hibernation. Some small animals, bats and oth-ers, retreat into daily torpor, thus avoiding the energetic expense of maintained metabolic rates while at rest. Desert-dwelling small mam-mals respond to food scarcity and daytime heat by withdrawing into a modestly depressed state of estivation within burrow enclosures. A few bird species maintain energetic stability with limited food gath-ering by conditions of nightly torpor lasting several hours (figure 7.1; Sharbaugh 2001).

Differing categories of torpor may be subject to similar modes of regulation dependent upon governing mechanisms originating in the central nervous system. The onset of hibernation is accompanied by a decline in metabolic rate that precedes the fall in body temperature. Respiratory depression results in gradual increase in carbon dioxide retention and associated increase in acidosis (Snapp and Heller 1981; Malan 1988, 1999). The sequence is reversed during arousal from hibernation.

The mammalian hibernator prepares for that metabolic retreat by several physiological adjustments. Its first line of defense against unfavorable seasonal extremes involves accumulating a caloric reserve in the form of body fat or by sequestering food for later consumption during arousal intervals. The energy-conserving advantage of hibernation is especially important for survival of small mammals in challenging seasonal environments. Without its metabolic savings their small body size would dictate a large surface-area-to-body-mass relationship and a consequent high rate of heat loss and elevated metabolic rate with its associated requirements for high food intake. The reduced energetic budget for the hibernating condition, including that required for regular arousals, can amount to a reduction of 90 percent or more of the value that would maintain the small animal in the nonhibernating condition. Bears, in contrast, have considerable capacity for onboard energy storage, and they prepare for long continuous hibernation by massive accumulation of enough body fat to sustain them for several months.

Ground squirrels and marmots (*Marmota caligata*) interrupt their hibernating state every few days, recovering near-normal body temperature for several hours, much of that time spent asleep, before renewed body cooling and return to hibernation. Reasons for this behavior, expensive in rewarming effort, have been the subject of speculation related to possible sleep deprivation during hibernation (Daan, Barnes, and Strijkstra 1991).

Hibernating black bears (figure 7.2; *Ursus americanus*) have adult body weight of 100 kg or more. Minimum oxygen consumption was determined during recording episodes, using an average of three bears. They experience a roughly continuous 75 percent reduction in metabolism during hibernation, despite a decline of no more than a few degrees in body temperature. That condition is related to their relatively reduced rate of heat loss and consequent slow temperature change of their large body mass with its superior insulation (Tøien

7.2.   Hibernating black bear (*Ursus americanus*). Drawing by Wendy Elsner.

et al. 2011). They maintain their hibernating metabolism at about one-quarter of its normal rate, sometimes varying that level but only rarely arousing. Their responses therefore markedly contrast with the thermal and metabolic alterations of the much smaller hibernators such as ground squirrels.

The precise mechanisms governing the onset of and arousal from mammalian hibernation are not well understood, and the search for an induction process or substance remains elusive. Success as a hibernator depends on the ability to adjust to the resulting drastic modifications of function. The fundamental mechanisms by which it is initiated and maintained frustrate detailed clarification despite vigorous and dedicated research and a resulting abundant literature (Heldmaier, Ortmann, and Elvert 2004).

Hibernating mode

The duration of torpid periods is highly variable, some lasting only a few hours: diurnal in bats, nocturnal in birds, days to weeks for ground squirrels and marmots, months for bears. Brief diurnal or nocturnal torpor is limited to a few hours of relatively modest response compared with the deep multiday withdrawals of hibernation. Arousals and rewarming occur in regular and predictable periods sometimes occupied with sleep and terminated by return to hibernation.

The dormant animal's reactions include several differing modes of response varying with species, body size, and environment, daily or nocturnal torpor, and winter hibernation. In contrast, estivation represents a similar metabolic depression that may be induced in some species during exposure to hot, dry climates in which a limitation of food availability may be a precipitating cause. Reduced metabolism is characteristic of these withdrawn states (Wang and Wolowyk 1988; Boyer and Barnes 1999; Geiser 2004). In Arctic ground squirrels this amounts to about 1 percent of the nonhibernating resting metabolic rate (Barnes 1989), contrasting with the bear's much higher hibernating metabolism. Body fattening and food storage, varying in different species, are the characteristic preludes in anticipation of the hibernating period. The animal then withdraws into a secure and insulated habitat appropriate for enduring the times of limited food resource until environmental conditions become more hospitable.

Body temperature does not free-fall during hibernation, but it continues to be readjusted around the depressed level, responding to variations in environmental temperature. The decline of metabolic rate precedes the lowering of body temperature by a brief interval. Maintenance of the low temperature in the heart and central nervous system is well tolerated in contrast to the threatened vulnerability of similar cooling in nonhibernators. The hibernating animal's responses are much reduced but not abandoned, and it may be aroused if it is excessively disturbed. Heart rate declines to much-reduced levels, to a few beats per minute in ground squirrels and from about fifty-five to fifteen beats per minute in black bears.

Deep body temperature of hibernating Arctic ground squirrels declines to near freezing, sometimes even supercooling below that level (Barnes 1989; Tøien et al. 2011), but it continues to be regulated at the reduced level (Heller and Colliver 1974; Florant, Turner, and Heller 1978; Boyer and Barnes 1999). It declines from the normal 37°C, even to negative and supercooling value (Barnes 1989). Its overall energetic savings amounts to about 95% of the nonhibernating metabolic rate. Heart rate slows from over 200 to less than 5 beats per minute. Brain circulation decreases to levels that would result in damage from lack of metabolic support in nonhibernators. Breathing becomes irregular and interrupted, sometimes ceasing entirely for several minutes. Arousals may occur periodically every one to three weeks. They recover sleep during these intervals.

Black bears, as noted, differ from ground squirrels and other hibernators in their considerably greater body mass and a characteristically more shallow metabolic depression and more moderately lowered body temperature, conditions related to their greater thermal inertia. They remain in a relatively subdued state in contrast with the regular awakenings of hibernating ground squirrels and other small mammal hibernators, maintaining a relatively steady metabolic rate. The hibernating black bear's metabolic rate, shown as oxygen consumption, and deep body temperature are shown in figure 7.3. These conditions have led to a suggestion that they are not true hibernators, but that they engage in a shallower metabolic condition. However, the bear's lower thermal conductance is evidence that its relatively modest reductions in temperature and metabolism are appropriate for its large body size.

Body temperature and food restriction
Temperature differences normally exist throughout the mammalian organism as a result of regional differences in heat loss and in circulatory distribution of warm blood. The effect is noticeable in both large and small mammals.

There are clinical advantages in modest hypothermia of 15°C to 30°C for surgical procedures in which the reduced metabolic activity allows sufficient time for procedures such as coronary artery bypass interventions. Attempts to simulate the extreme body cooling in nonhibernating species are rarely successful, but hypothermic survival is enhanced if combined with hypoxia (Wood and Gonzales 1996). Its virtue as an adaptive response is based on the reduction of oxygen requirement associated with decreased body temperature. These conditions result in modest metabolic modifications for small temperature changes close to the normal body value, but they represent exponential functions and increase considerably with body temperature changes typical of some hibernators.

Hypothermia, as it occurs naturally in animals by virtue of a decline

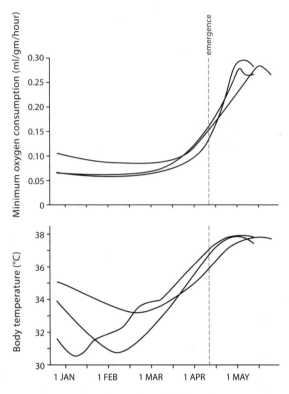

7.3. Minimum oxygen consumption and core body temperatures of bears during and emergence from hibernation. Individual values of three black bears. Redrawn from Tøien et al. 2011.

in body temperature consequent to cold exposure, is accompanied by a reduction in blood pH and modestly increasing acidic conditions. It is recognition of this response that requires attention to adjustment of pH for the successful use of hypothermic surgery (White 1981; Kroncke et al. 1986). Reduced body temperature is a routine response to hypoxic exposure that occurs in mammals of a variety of taxonomic groups (Wood and Gonzales 1996). Lowering of metabolism and body temperature also occurs in some mammalian species in response to hypoxia, especially those of small body mass (Frappell et al. 1992). Similar effects are seen in nonhibernators of varied taxonomic identities when reacting to restriction of food intake (Walford and Spindler 1997). Metabolism is modestly suppressed in some large animals, Shetland ponies, for example, when food is limited (Brinkmann, Gerken, and Riek 2012).

The central nervous system of the Arctic ground squirrel is unusual in its tolerance of reduced blood flow in the euthermic state, a condition perhaps related to vulnerability during initiation of and arousal from hibernation (Drew et al. 2001, 2007; Dave et al. 2006). Similar beneficial effects have been shown in some hibernators' innate resistance to cardiac ventricular fibrillation (Johansson 1996). Brain cells of ground squirrels are unusual in their tolerance of oxygen deprivation in the nonhibernating condition (Dave et al. 2006; Drew et al. 2001). This characteristic resistance to injury has been an incentive for pursuit of an intrinsic chemical substance that may trigger the onset of torpor.

The prospect that a chemical mode of hibernation induction may be identified in blood of natural hibernators has attracted attention. Search for such a chemical mechanism active in initiating natural hibernation has not been entirely successful and continues, but isolation of specific substances believed to be directly involved has led to prospects of direct chemical induction. Possible use of such a substance as an adjunct to surgical procedures, especially those requiring cessation of cardiac contractions, has led to successful experimentation with mimetics that have properties of torpor induction in mammalian species, both hibernators and nonhibernators (Bolling et al. 1997; Husted et al. 2004).

Human reactions

Humans are among those species for which recourse to hibernation is not an alternative, although transient reductions of specific organ metabolism can occur. Among these are the effects of depressed cardiac function in humans: myocardial "stunning" (Patel et al. 1988) and "hibernating" (Rahimtoola 1989). These conditions refer to the reduced metabolic state of the heart during slow recovery from some persisting pathological conditions.

There are clinical advantages in temporarily induced hypothermia to deep body temperatures of 25°C to 32°C for cardiac surgical interventions in which the reduced body metabolic activity allows sufficient time for procedures such as that required for coronary artery bypass in treatment for cardiac disease. Attempts to induce prolonged extreme body cooling in nonhibernating species are rarely successful, but hypothermic survival may be enhanced if combined with hypoxia (Wood and Gonzales 1996).

Complete circulatory arrest of blood flow in human heart muscle results in prompt elimination of cardiac contraction and pumping action. If normal perfusion is restarted before necrotic deterioration has begun, heart function may be revived. These are clinically recognized conditions from which the heart recovers slowly. Similar reactions occur in the hearts of other mammals, but the extent and tolerance of trauma and the rapidity of recovery vary. The seal heart tolerates severe and prolonged asphyxial hypoxia far better than do the hearts of most terrestrial mammals (Elsner et al. 1985; White et al. 1990).

Regulatory mechanisms, hibernators, and divers

An important mechanism for metabolic suppression in hibernators relates to their ability to reduce respiration. The consequent carbon dioxide retention and inhibition of enzymatic activity lead to an associated increase in acidic condition (Bharma and Milsom 1993; Malan 1999). In these respects their reactions resemble those of diving mammals but with a different time course, that is, the diver's rapid onset and relatively brief duration, minutes to an hour or more in some diving mammals, contrasting with days and months in hibernators. A common feature is respiratory suppression leading to carbon dioxide retention and the consequent cellular acidosis (Heller 1988). The mechanisms by which metabolic suppressions in hibernation and diving are induced and regulated are therefore useful topics for comparative consideration.

Research efforts have directed considerable attention to the means by which hibernation is regulated by the central nervous system. A likely function of the suprachiasmatic nucleus of the anterior hypothalamus in the brain stem has been suggested as both a rate-setting pacemaker and a regulator of energy balance with likely implications for its role in hibernation. Throughout deep hibernation in ground squirrels, the excitatory activity of the central nervous system is suppressed into a quiescent state, with the notable exception of the suprachiasmatic nucleus, which continues to show electrophysical activity, finally ceasing below about 12°C (reviews: Heller and Ruby 2004; Ruby 2003).

The reactions of the organism during the onset of hibernation resemble in some respects those accompanying the onset of sleep. Similarities of central nervous system activity at the initiation of torpor and sleep have been noted (Heller and Glotzbach 1977; Florant, Turner, and Heller 1978; Heller 1988). Both conditions are characterized by suppression of thermoregulatory responses and the shift to a new and lower level of body temperature (Berger and Phillips 1988).

The issue of circadian rhythmicity during hibernation has attracted considerable attention. Its characteristic display appears in animals engaged in daily torpor in which body temperature declines are relatively modest, but its level of activity is reduced or absent in animals committed to longer hibernation with lower body temperature. These thermal rhythms are abolished with ablation of the suprachiasmatic nucleus of the brain, and the subsequent duration of hibernation is considerably extended by this procedure in some animals (Heller and Ruby 2004).

The association of diving with control functions of brain centers has received less attention than reactions of the more peripheral neural circuitry elements. There are, however, preliminary observations that may lead to promising research of some interest. Recordings from the circulation in an elephant seal suggest an association of the cardiovascular reactions to immersion with electrical stimulation of the periventricular anterior hypothalamus, the same region within which the suprachiasmatic nucleus is located and active in the hibernating mammal. Responses of arterial pressure and regional blood flow were similar to those recorded from the animal during submersion (Van Citters et al. 1965). Electrical activation of other cortical and subcortical regions of the seal's brain did not result in similar cardiovascular effects.

The more peripheral neural components of the diving response are better understood, due in large part to research efforts of Michael de Burgh Daly and his colleagues as described in publications from his laboratory and in his monograph (Daly 1997). They have shown that the regulatory responses to underwater submersion are not unique to those animals that routinely engage in diving. Rather, they are modifications of the processes that govern respiration and metabolism and their interactions in animals. Their common cardiovascular reactions to immersion and cessation of breathing are the responses of bradycardia and peripheral vasoconstriction. Immediately upon breath-holding submersion, heart rate slows, often abruptly. This cardiac effect is followed by a decline in blood and cellular oxygen. Brief immersion results in little or no disturbance of the equilibrium state; longer dives result in more disruption and longer recovery. These conditions are characteristic of vertebrate divers but with many variations depending upon species.

As the dive lengthens, blood levels of lactate and hydrogen ions appear in increasing concentrations in blood and extracellular fluid. The progressive increase in hypoxia, hypercapnia, and acidosis stimulates chemoreceptors, primarily in the carotid bodies located at the junction of internal and external carotid arteries in the neck. Their activation, normally resulting in increased breathing, is suppressed by the persisting stimulation from sensory innervation of the face and airway responding to the presence of water (Daly, Elsner, and Angell-James 1977; Elsner and Daly 1988). In brief dives these stimuli are insignificant, but they become increasingly active as the dive progresses. These reactions are subject to considerable modification by the influence of higher brain centers; fear or escape, for example, leads to enhancement of the defensive reactions and reduction of diving endurance.

The tolerance of the seal's brain and heart for exposure to comparably low arterial oxygen levels, along with modest reduction of body temperature in long dives, bears a resemblance to the hibernator's endurance of hypoxia. Chronic high-altitude exposure of mammals to low atmospheric oxygen also results in development of cardiac resistance to hypoxia (Poupa et al. 1966; Ostadal and Kolar 2007) and tolerance of cerebral structures to similar exposure (Hochachka, Rupert, and Monge 1999).

The hibernating turtle

The overwintering condition of turtles and other "cold-blooded" ectotherms is often referred to as hibernation. Their reactions to season-

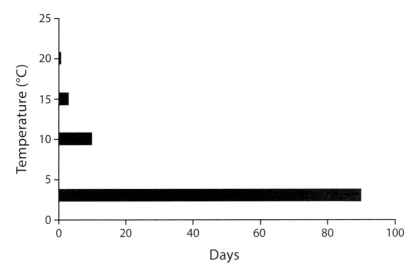

7.4. Hibernating turtle (*Chrysemys picta belli*), duration varying with environmental temperature. Redrawn from Jackson 2000.

ally reduced metabolism are highly variable. Hibernating freshwater turtles remain underwater, often buried in bottom mud, for several months (figure 7.4). Several freshwater turtle species have been studied as representative air-breathing vertebrates that are tolerant of oxygen lack such as can be expected to exist in such conditions (Jackson 2000; Wasser 1996). The turtle's resistance to these adverse effects is based on an ability to endure profound reduction of metabolism to about 10 percent of the normal rate that would function at comparable temperatures. It also resists the hazard from increased acid metabolic products by neutralization in the buffer reserves of its shell (Jackson 2000). Its ectothermic metabolic rate is roughly one-fifth to one-tenth that of a similar-sized endotherm. Accompanying conversion to anaerobic metabolism lowers the metabolic rate to about 10 percent of its aerobic rate. For the turtle at near-freezing temperature, these factors combined yield a metabolic rate that is only 0.5 percent of the turtle's rate at 20°C (Jackson 2000). This value amounts to less than 0.1 percent of the metabolism of a similar-sized mammal under comparable conditions. The reduction in cardiac activity may be indicated by an observed rate of sometimes no more than one beat in ten minutes at 3°C.

The ability of some turtle species to tolerate long and severe experimental anoxia is accounted for by their metabolic depression to a small fraction of normal. This condition is supported by their abundant glycogen stores in liver and brain for anaerobic metabolism (Lutz, Nilsson, and Pérez-Pinzón 1996) and their resistance to the unfavorable effects of accumulated acidic metabolic products by the buffering of these substances within the turtle's shell (Jackson 2000, 2004). These resources and mechanisms support the turtle's survival during long submergence and while it remains winter-dormant in pond mud. Metabolic conversion to anaerobic processes, amounting to only a small fraction of that obtainable from oxidative phosphorylation, and the dependence of turtle brain on a considerable reduction of energetic exchange account for its long-term survival in this environment.

Freshwater turtles (Jackson 2004), frogs (Donohoe and Boutilier 1998), snakes (Hermes-Lima and Storey 1993), snails (Hermes-Lima, Storey, and Storey 1998; Storey and Storey 1990), and insects (Zachariassen 1985) experience thermally driven seasonal metabolic inhibition. Their apparent suspended animation is a condition in which the minimum signs of life are absent but from which revival is possible (Boutilier 2001). Bacterial spores and plant seeds can exist in a desiccated state for years or centuries. Some are reported to have persisted in a dormant state for 250 million years and have been revived with appropriate treatment (Vreeland, Rosenzweig, and Powers 2000).

Manipulating metabolism

A general response to induced hypoxia is often noted as a reduction in metabolism. It is especially noticeable in lower vertebrates and newborn mammals.

The idea that it may be possible to induce a protracted state of extreme metabolic depression with modest reduction of body temperature, even in humans, has captured the imagination of science fiction writers for many years. The reality is nevertheless provocative, as suggested in reports of survival from severe accidental hypothermia (Kornberger and Mair 1996). Profound hypothermia of 10°C has been successfully induced in dogs for one to two hours by rapid temporary replacement of blood with cold saline (Behringer et al. 2003; Wu et al. 2006).

Recent studies by Roth and his colleagues have shown that in certain conditions of near-anoxia in which oxidative phosphorylation ceases, a reversible state resembling suspended animation can be induced in embryonic nematodes by introduction of carbon monoxide (Nystul and Roth 2004). A similar state is inducible in mice in the presence of hydrogen sulfide (Blackstone, Morrison, and Roth, 2005). Roth and Nystul (2005) describe the condition as "survival of the slowest." The decrease of available oxygen to low levels ordinarily results in cellular disruption leading to death. In contrast, however, total elimination of life-supporting oxidative reactions can in the tested organisms lead to a state of suspended animation in which metabolic processes characteristic of life are suppressed, yet the organism survives.

An unexpected finding of this research is the determination that, despite the potential poisonous attributes of carbon monoxide and hydrogen sulfide, the consequence of administering these metabolic poisons in blocking or competing with oxygen uptake can under conditions of severe hypoxia actually preserve life by imposing suspended animation. The technique has been successfully applied to several organisms from a wide range of species including nematodes, zebrafish, and mice. It represents a novel approach that is likely to reveal new and unexpected considerations of what we regard as the limits of our capabilities for manipulating life.

Application to humans of similar techniques for manipulation of body temperature and metabolism has been suggested as a means for treatment of casualties resulting from severe accidents and battlefield wounds. The goal is the management of ordinarily untreatable massive blood loss and disruption of internal organs of the chest and abdomen, allowing sufficient time for transport to more complex and suitable treatment facilities. Success of this concept would depend upon rapid induction of a state of suspended animation in which brain function is preserved while respiratory and cardiovascular activities are rendered inactive.

HUMAN

8

DIVERS

We do not ordinarily regard people as diving mammals, but human breath-holding skin divers have a long history of underwater work and food gathering from the sea. While engaged in these activities, humans experience physiological reactions resembling in some respects those of our aquatic relatives, identifying us, however modestly, as part of a diving-mammal continuum. Our underwater endurance is obviously far removed from that of seals; nevertheless, human responses representing asphyxia defense reactions bear a distinct but much-diminished resemblance to the seal's superior ways for dealing with its customary habit of prolonged breath-holding.

Human breath-hold diving is widely practiced today as a popular recreational activity. It also persists as a regular occupation in Japan, Korea, Indonesia, and Polynesia, but it has declined generally as a profession in most regions of the world where it has been practiced for centuries. Skin divers, professional as well as recreational, usually limit the duration of their underwater excursions to not much longer than a minute and descents of usually less than twenty meters depth.

The few extraordinary examples of long breath-holding extremes do, however, reveal an unexpected range of this human physiological capacity. Perhaps the deepest routine breath-hold dives of contemporary times are those occurring in recent years in Polynesia. Breath-holding pearl divers of the Tuamotu Islands have been known to regularly dive to depths of forty meters (Cross 1965). These unusual human examples derive from centuries-long traditional occupations, some mentioned in ancient literature. The earliest may be that in Homer's *Iliad*, about 800 BC.

Various aspects of human breath-holding immersion in water have been examined by several investigators in recent decades. Despite our modest ability for coping with respiratory disruption, some unexpectedly long accidental submersions have been recorded. One might be tempted to attribute these unusual endurances to an unexpected extension of the seal's diving adaptations, but that is unlikely to be an appropriate explanation. Survival may more reasonably depend upon other effects not strictly related to the diving responses as we ordinarily understand them. It may more likely be the result of rapid body cooling, thereby lowering metabolism and thus protecting the highly vulnerable central nervous system. This fortuitous outcome may be more readily achieved in small children than in adults, because of their relatively large ratio of surface area to body volume, thus resulting in more rapid heat loss to the water. The consequent decreased need for respiratory support, most critically by the central nervous system, may thus extend their brief limits of asphyxial endurance.

The professional breath-hold divers show only modest similarity to the seal's diving adaptations. This is not to suggest that their diving skills are trivial; they are, however, the product of a highly disciplined training regime that results in efficient use of underwater effort for maximum productivity. The results of heart rate determinations from more than a thousand human experiments of breath-holding face immersion have been summarized by Arnold (1985). Nearly all human breath-holders show some degree of heart rate slowing, an effect greater during immersion than when breath-holding in air. The reaction is sometimes surprising. Approximately 2 percent of the subjects responded with intermittently low heart rates under thirty beats per minute; a few even fell to ten to twenty beats per minute (review: Elsner and Gooden 1983) and an extreme low of nine beats per minute (Arnold 1985)!

These extraordinarily depressed heart rates were well tolerated without fainting or other adverse reactions, suggesting that adequate cerebral blood flow was maintained by compensatory peripheral vasoconstriction. In other tests, limb blood flow was reduced in young men during breath-holding face immersion, a few with near-complete cessation of flow (Elsner et al. 1966). The responses are consistent with a general cardiovascular response to apnea and immersion. Compared with seal dives, they suggest that modest or negligible oxygen conservation results from these reactions in humans. Whatever beneficial effects there are may be limited to similar responses in threatened fetal and newborn animals that can sometimes be life-sustaining.

The performances and physiological reactions of Japanese and Korean breath-hold divers, food and shell gatherers, have been subjects of international attention and research. Japanese *ama* divers were exclusively women in former years, a professional preference attributed to their better tolerance of cold exposure resulting from the frequent immersions of their working schedules. Their repeated working exposures resulted in metabolic and circulatory reactions of rare human cold adaptation (Hong 1963, 1973). That routine exposure was much reduced a few decades ago with the more customary wearing of protective rubber wet suits that came into general use at that time. Descriptions of the ama diver's routine are provided by Hong and Rahn (1967).

When the review volume edited by Rahn (1965) was published, about sixteen thousand women and men were engaged in this occupation in Japan, another three thousand in Korea. They were supported by the high market price of their products, mostly marine invertebrates such as abalone and snails. Some have been employed in the maintenance of cultured pearl operations. These free-diving activities continue today but with fewer participants, as other less arduous

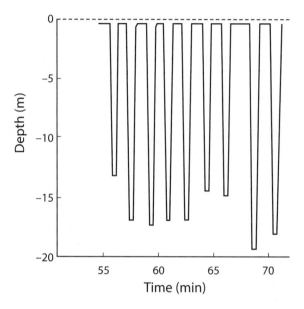

8.1. Diving depths, frequency and durations of Japanese funado diver. Redrawn from Mohri et al. 1995.

and more rewarding work opportunities have become available. The number of registered Japanese ama divers is currently reported to be much reduced and to be composed of more men than women (K. Shiraki, personal communication), indicating the gradual change and numerical decline in this profession. Several are said to be over eighty years of age; few new recruits are joining the profession.

The diving schedule of Japanese ama consists of a typical three- or four-hour working day during which up to one hundred dives are performed (figure 8.1). Dive depths range from four to twenty meters. Both *cachido* (shallow) and *funado* (deep) divers breath-hold about a minute, occasionally more, in each dive (Park, Shiraki, and Hong 1983; Mohri et al. 1995). Some ama use an inflated automobile inner tube for a free-floating station to which they return with their retrieved products from the sea floor. The relatively few deeper funado divers work from a skiff, their submergence assisted by a weight tied to a rope and their retrieval to the surface boat by an assistant (figure 8.2). This procedure reduces swimming effort and provides for efficient working time at the bottom.

8.2.   Funado diver at work. Redrawn from Kita 1965.

Physiological reactions

Although human divers suffer by comparison with marine mammals, their reactions show resemblances that suggest similar physiological effects. The performance of ama divers depends principally on abilities learned by long practice rather than on special physiological attributes. Their adjustments to repeated breath-holding and cold exposure are nevertheless readily detectable. Decompression sickness (*bends*) resulting from repetitive deep dives is not reported to be a problem for ama divers, probably as a consequence of their extended surface intervals for recovery between dives. This cautious schedule likely prevents the absorption of excessive nitrogen in tissues in which potentially hazardous bubble formation of decompression sickness might develop. Some suggestive evidence of incipient bends was reported to result from the deep dive schedules of the Tuamotu islanders (Cross 1965).

A determined effort to reproduce experimentally the maximum efforts of deep breath-hold dives resulted in decompression symptoms (Paulev 1965). About sixty rapidly repeated dives to fifteen to twenty meters were performed in a submarine escape tank, these involving breath-holding for up to two minutes in each dive. After continuing this demanding schedule for five hours, the diver complained of typical bends symptoms, joint discomfort, nausea, and dizziness, all of which were promptly relieved by recompression in a pressure chamber. The effects suggest that decompression sickness can be induced by severely pushing the depth and frequency limits of voluntary diving. Whether or not the lifelong diving experience of the ama provides some protective adaptation against bends remains a matter of conjecture. Routine exposures of funado divers to deeper submersions than those of cachidos results in increased compression of their pulmonary gases and consequent increased blood partial pressures, although their dive frequency is less. The funado divers were found to have a reduced sensitivity to the generally increased respiratory stimulation of breathing in reaction to their elevated concentrations of pulmonary carbon dioxide (Masuda et al. 1982). A similar effect appears in US Navy submarine escape trainers who experience frequent free ascents from depth (Schaefer 1965), a condition that is gradually reversed in the months following termination of that activity.

The ama diver's primary recognizable physiological adaptation is acclimatization to cold exposure, or it was before they were better protected by wet suits, beginning about 1977. A consequence of extended periods of frequent diving before the general use of protective wet suits was their obligatory exposure to the body cooling associated with long periods of immersion in cold water. Japanese ama diving operations are generally limited to the warmer months of the year, but Korean divers have been regularly subjected to cold exposure induced by their continued working in winter months. Before the introduction of wet suits, the divers worked for shorter periods until they became uncomfortably chilled, then returned to the shore for rewarming by an open fire. The duration of diving episodes was found to be dictated by body cooling sufficient to lower deep body temperatures of the divers to about 35°C. When the physiological studies of Korean divers were initiated by Suk Ki Hong and his coworkers in 1959, winter dives involved unprotected routine exposure to water as cold as 10°C, and diving work was limited to brief periods separated by shoreside rewarming intervals. The divers have been found to be relatively lean, as estimated by subcutaneous fat thickness. Body temperature declines of 2°C or more were reported to be routinely detected during diving activities (Rennie 1965).

The issue of human physiological adaptation or seasonal acclimatization to cold water exposure has been a matter of conjecture for a half century or more, and the Korean women divers have provided a model population for its testing. Protective rubber wet suits and swim fins have been widely used in both Korea and Japan as a necessary condition of the diving occupation. It is fortuitous, therefore, that the modern studies of Korean divers were initiated by Hong and Rennie and their colleagues (Hong 1963, 1973; Rennie 1965; Hong and Rahn 1967) before the routine use of insulated protective clothing. This recourse to improved protection permitted a much less severe cold exposure and a lengthening of diving work schedules. The unusual opportunity provided by this major change in the diving environment was the rationale for an extensive series of before-and-after studies that have clarified the severity of physiological impacts relating to the diver's working conditions. They were tested by determining reactions to cold water, critical water temperature, defined as the temperature that could be tolerated for three hours without shivering, and the related increase in oxygen consumption (figure 8.3). This relatively steady-state condition results in maximum tissue insulation at water temperatures ranging between 29°C and 33°C, depending on the subject's natural insulation derived from the thickness of subcutaneous fat and peripheral tissue.

Before general use of protective wet suits, the Korean divers were found to experience a seasonal increase in basal metabolic rate. They tolerated comfortably a critical water temperature well below that of both Korean nondivers and comparable American controls despite their relatively thin deposits of subcutaneous fat (Rennie 1965; Kang et al. 1963; Hong 1973; Hong, Rennie, and Park 1986). The elevation of metabolic rate amounted to a 12 percent increase above that of nondivers recorded during the winter diving season. That this metabolic response was an effect resulting from frequent exposure to cold water appeared to be confirmed by its disappearance noted a few years later when insulated diving suits came into wide use and by its contrasting absence in neighboring nondiving individuals matched for age and other characteristics. It represents one of the few clearly

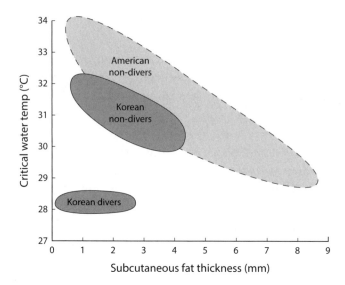

8.3. Critical water temperature (coldest tolerated for three hours without shivering) and subcutaneous fat of Korean ama divers, nondiving Korean women, and American nondiving men and women. Redrawn from Rennie 1965.

recognized and described instances of human metabolic adaptation to regular cold exposure. However unusual this response to cold may be, it represents at best no more than a relatively modest, albeit readily verifiable, contribution to the diver's overall metabolic rate. Hong's (1965) appraisal is pertinent: "Although the observed increase in the basal metabolic rate of the Ama in winter has physiological significance, the practical value must be minor as a defense against body cooling in winter."

Human metabolic responses tending toward improved function or protection have been difficult to verify in purposeful exposures of nondivers. An attempt was made to experimentally match the exposures of Korean divers in three volunteer Caucasian women subjects (Rennie, personal communication). The subjects were cold-exposed in water at 21°C for one hour daily from November to March. Despite their best heroic efforts, they found the exposure barely tolerable and were unable to match or even to approach the cold temperatures of the Korean divers' work environment. One of the subjects showed a modest increase in basal metabolism, but further experimental ex-

posures proved to be too uncomfortable, even for these highly moti-
vated subjects. In contrast to the uncertain nature of metabolic adaptation to cold, there is general agreement that local responses to cold,
such as increased peripheral circulation resulting in improved hand
dexterity and appendage comfort, are readily achieved by repeated
exposure extending over days or weeks.

Extraordinary exposures

Among the examples of severe cold water exposures, a few instances
of extraordinary survival have been recorded. The possible high rate
of survival failures in these circumstances may be less likely to be
noted and reported. Successful recoveries from long submersions are
often those of small children in cold water. Cold immersion and related metabolic depression can sometimes be more protective than
warm immersion, and successful recovery may depend upon the
simple principle that small bodies cool faster than large ones. Examples of such survival, two children and three adults, are worth noting in some detail.

The immersion of a five-year old boy was recorded as lasting forty
minutes after he broke through the ice of a river (Siebke et al. 1975).
His body was finally located by rescue personnel and was recovered
from about three meters depth. Mouth-to-mouth emergency respiratory support was started immediately and continued while he was
being rushed to a hospital, arriving there about an hour after immersion. He showed no signs of spontaneous respiration or circulation; his
eyes widely dilated, and his rectal temperature was 24°C.

External cardiopulmonary resuscitation was administered in addition to intravenous sodium bicarbonate in an effort to correct acid-
base imbalance. Plastic bags containing warm water were applied to
his body for improvised rewarming. Cardiac activity was not detected
until an hour after admission to the hospital. Both urine and plasma
showed signs of hemolysis, suggesting that some water had been transferred across the pulmonary membranes of the lungs and come into
contact with circulating blood. Spontaneous circulation was eventually detected, and rectal temperature reached 37°C eight hours after
admission to the hospital. His recovery then proceeded rapidly, and

he was discharged from the hospital eight days after the accident. Neurological symptoms, such as memory loss and poor coordination, persisted for several days. On examination four months later he was found to be further improved, with only a slightly disturbed postural coordination remaining. Apparently normal physical and intellectual function was evident a year later.

Adult recoveries from drowning conditions appear to be less successful than those of children. Despite the usual adverse consequences of severely cold immersion, some notable exceptions have been reported, and the occasional occurrence of recovery from accidental hypothermia provides some insight into the extremes of human body cooling tolerance. One of the more remarkable examples is that of a twenty-nine-year-old Norwegian woman who survived long accidental immersion in cold water resulting from an unusual skiing accident. The victim, an expert backcountry skier, fell through snow and ice covering a cascading river and was inescapably trapped and immersed in such a way that neither she nor her companions could free her. She struggled for several minutes and was then quiet and presumably unconscious. The precise duration of her immersion is uncertain, but it was reported to have been a minimum of forty minutes.

She was finally extricated, lifeless in appearance, by a rescue team about an hour after the accident. Transported to hospital by helicopter, she was found to have a rectal temperature of 14.4°C, declining shortly thereafter to 13.7°C, and appeared to be clinically dead. Cardiopulmonary bypass and resuscitation with warming were initiated and continued for one and a half hours. By that time the patient's rectal temperature was 36°C. Resuscitation continued for nine hours, maintained despite several complications. She was then transferred to intensive care, where she remained on assisted ventilation for thirty-five days. This initial treatment was followed by several weeks of continued management and hospitalization. Five months after the accident she had recovered sufficiently to return to her normal activities and to the resumption of recreational skiing (Gilbert et al. 2000).

Two instances of adult recoveries from long cold water immersion, both in the Charles River, Boston, were reported by Sekar et al. (1980). A twenty-three-year-old woman was immersed for 25.5 minutes. She appeared lifeless upon arrival at the nearby hospital except for a detectable heart rate of 35 by ECG, with an arterial blood pH of 6.83 and a rectal temperature of 28.8°C. Rewarming was accomplished during the subsequent several hours, followed by vigorous treatment during the next several days. She was discharged from the hospital after fifteen days and continued a steady recovery and return to apparent good health. The other patient was a quadriplegic twenty-seven-year-old male who drove his motorized wheelchair into the river. He was rescued after six minutes of immersion, with a rectal temperature of 33°C and arterial blood pH of 7.24 at the hospital. He was discharged, recovered, seven days later.

An extraordinary survival was that of a child submerged in cold water for sixty-six minutes. This well-documented account describes the sequence of treatments administered. Initial efforts at cardiopulmonary resuscitation failed after being continued for more than two hours. Ultimate recovery followed extracorporeal rewarming from a deep body temperature of 19°C. The report of this successful procedure includes a review of treatment options and prospects for success in similar recoveries from accidental immersions (Bolte et al. 1988).

A few other seemingly miraculous recoveries from drowning have been recorded (Elsner and Gooden 1983). Requirements for successful resuscitation in similar situations depend usually upon the victim's young age, cold water temperature, and aggressive treatment. It may be tempting to suggest that an exaggerated diving response, protective of the central nervous system, is a contributing factor. Such an effect is probably restricted to no more than the first minutes of immersion. That condition, if it occurs, might help to limit or prevent the aspiration of water, but the prospect for long-term recovery is more likely to depend upon depressed metabolism, a consequence of rapid body cooling.

These and other instances of successful outcome from life-threatening exposures raise questions regarding the reactions that occur in such extraordinary situations (Gooden 1992). Recovery from these accidents is likely attributable in large part to the associated cooling and consequently depressed metabolism, especially that of the central nervous system. Survival of the Norwegian woman described might have depended on her rapid cooling by continuous immersion in near-freezing flowing water. Whatever role a functioning diving

response played in the recovery process would most likely be con-
fined to the immediate effects in the first few minutes of immersion,
perhaps resulting in minimal protection before central nervous system cooling had occurred. The examples of successful recovery from
long submersions should not obscure the reality of frequent survival
failure in severe instances for which humans are poorly endowed.
Accidental and lengthy cold exposure, in water or air, is highly threatening to life, and successful recovery is uncertain at best (Walpoth and
Daanen 2006).

Human limits

When compared with the clear examples of physiological modification imposed by long-term or repeated exposures to diverse environments, the prospects for practiced enhancement of the human
dive duration and depth of breath-holders are severely limited. We
recognize the limit of our comfortable breath-holding tolerance at
about a minute. Human breath-hold divers of vast experience, professional Japanese and Korean divers are not especially noted for unusual breath-holding endurance and dive depths. The skills produced
by their training are primarily those of keen sensory perception and
accommodation to their underwater harvest work, less by increased
tolerance for enduring long breath-holding.

Lin (1990) considers various physiological limitations that restrict
maximum expressions of breath-hold diving. The instances of such
competitive dives to a hundred meters and similar submersions lasting
five minutes (Schagatay, Richardson, and Lodin-Sundström 2012) are
rare events performed by individuals who have made extraordinary
efforts for their maximum expression. The reported current human
world record breath-hold time is 8 minutes 47 seconds (Schagatay,
Haughey, and Reimers 2005)! Such an extraordinary departure from
ordinary capabilities is the apparent result of a combination of innate
capacity, diligent training, and preparatory long-duration deep breathing and breath-holding. This preliminary increased rate and depth of
breathing contributes importantly to breath-holding endurance by the
accompanying lowering of carbon dioxide blood and tissue content
and the resulting depression of respiratory stimulus. Some evidence

suggests that spleen contraction and the resulting increase in circulating blood red cells may contribute to the red cell mass available for oxygenation (Schagatay et al. 2001) and might by this effect support the lengthy breath-hold.

The environments that are tolerated by humans are in ordinary circumstances strictly limited to benign and modest conditions when viewed from the comparative perspective of what some of our mammalian relatives find endurable and maybe even comfortable. At least two aspects of frequently encountered exposures impose such severe limits on our activities that they are not ordinarily considered to be a part of life as we know it. These are cold exposure and breath-holding, conditions for which some other mammalian species are better equipped. Human physiological constraints impose relatively tropical and aerobic lifestyles; survival in most of the world as we know it is made possible only by living well clothed and housed in protective environmental isolation, the likely consequence of our warm-climate origins and physiological limitations.

Examination of human cold tolerance, such as may exist in a few human populations, is instructive about that limited endurance. A worldwide search for examining the reactions of native groups suspected of being historically cold-exposed was undertaken several decades ago. The studies were inspired by Scholander's concept that the populations to be examined might be expected to be most vulnerable during their normal time of sleep. Appropriate testing procedures were considered to include determinations of metabolic and thermal responses throughout the regular full night's sleeping period with standardized minimum protection that simulated or was less than their customary usage. These included metabolic rate (determined as oxygen consumption) and body temperatures, skin and rectal, recorded regularly throughout the night.

These physiological responses were compared with those of the research team participants, all of whom were known to lack chronic or recent cold exposure. The timing of these studies was fortuitous, because many of the world's native populations were beginning to experience substantial alterations and improvements of their lifestyles involving conversions to more protective and comfortable conditions. The study groups included Australian Aborigines, Canadian Eskimos,

Arctic Indians, Norwegian Lapps, Chilean Alakaluf, Kalahari Bush-men, and Peruvian high-altitude natives. Their history of cold exposure was verified by documentation of their histories and lifestyles (review: Hammel 1964).

The initial speculation that a general set of similar responses may exist among the various groups was dispelled by the unexpected recordings of differing reactions, generally consistent within each native group. Physiological reactions of these presumably cold-adapted native groups varied widely in response to their standardized testing in similar conditions.

A few generalizations were apparent. Those whose regular sleeping conditions exposed them to moderate cold (Australian Aborigines, Kalahari Bushmen, Peruvian highlanders) showed a gradual cooling of deep body and skin temperatures and declining metabolism. In contrast, Eskimos, Alakaluf, and Arctic Indians, living in more severely cold environments, showing higher resting oxygen consumption and maintained steady or elevated metabolic and thermal responses. The Caucasians slept fitfully and with increased oxygen consumption and lowering skin temperatures, contrasting with the relative comfort of native subjects during much of the night (Elsner 1963 Hammel 1964).

The reaction to cold exposure of the Australian aborigines studied was one of gradually reduced metabolism and lowered temperatures of both deep body and skin (Hammel et al. 1959). Their metabolic decline during sleep was greater than that of participants in corresponding studies of all other ethnic groups, but their metabolic rate fell no more than approximately 15 percent during eight hours of sleep, far less than the 40 percent decline in four hours of the meditating yogi in the Heller, Elsner, and Rao (1987) study described in chapter 3.

A generalization that appears from the results of these studies is that habitual cold exposure sufficient to pose a likely threat of cold injury is likely to result in a response preventing or reducing that possibility. It is suggested that a specific acclimatizing process can be anticipated to occur in the peripheral circulation, notably that of hands and feet. The resulting increased circulation of warmth to those appendages expands their capacity for continued function, digital manipulation, and comfort. Adaptive responses of the organism for coping with the

hypoxia of high altitude are another example of an integrated physiological adaptation to a potentially threatening environment. Increased
carrying capacity in pulmonary and circulatory systems during the
early days and weeks of exposure contributes to the animal's functional capacity for coping with the threat of reduced environmental
oxygen.

RESISTANCES

TO

ASPHYXIA

Operation of the mammalian brain requires nearly continuous metabolic support. Deprivation of this essential resource, briefly in terrestrial species, longer in the seal, exhausts its reserve capacity and leads
inevitably to functional collapse. The seal's characteristic adaptations
for long breath-holding provide a model of mammalian asphyxial endurance obviously greater than that of its terrestrial relatives. That
capacity is, as we have seen, similarly dependent upon a maintained
source of metabolic support, the seal's by enhanced provision of both
aerobic and anaerobic resources for diving endurance.

Understanding the details of whatever mechanisms might exist for
resisting threats to consciousness clearly deserves attention. Failure of
the brief resistance to asphyxia in human and other terrestrial mammalian central nervous systems is so precipitate and catastrophic that a
systematic examination of its process can be a difficult and uncertain

undertaking. Some of this difficulty is rendered manageable by examination of appropriate comparative species in which the course of deterioration occurs more slowly and its effects may be better observed and experimentally manipulated. This approach has been explored in the book by Lutz, Nilsson, and Prentice (2003) in their chapter headed "Clinical Perspectives":

"The basic argument for believing that such comparative investigations may provide useful clinical insights is that in the energy deprived mammal brain the degenerative events are so rapid and complex that it has proven extremely difficult to sort out cause and effect. In anoxia tolerant species, on the other hand, we have unique animal models where the control situation is survival instead of death. There appears to be wide agreement that identifying the key processes that are protected in anoxic tolerant animals, as against those that are allowed to decline, may indicate promising targets for successful clinical intervention. The main targets are the regulation of the controlled metabolic suppression and the mechanisms of preservation of cell structure and membrane functions and integrity under conditions of severely reduced energy supplies."

When the seal dives

The scope of the seal's reactions to interrupted breathing is demonstrated by the manner in which it copes with requirements of the diving lifestyle. Our steadily developing insights regarding the seal's adaptations have improved our ability to ask more pertinent questions of other species, ourselves included. This route to new knowledge of how seals and other marine mammals live is an example among many of what we can learn from comparative studies that deal with responses of differing species to similar environmental exposures and impacts. Satisfying our curiosity regarding the seal's underwater reactions is not only an essential step toward our appreciation of its physiology; it is also an opening into fuller comprehension of some reactions in other mammals, including that of the meditating human. Metabolic suppression of the extent indicated demonstrates, perhaps unexpectedly, that the meditative state may entail an unusual aspect of the human physiological repertoire. It suggests that metabolism may be deeply suppressed by the relatively simple and benign technique of meditation and that this reaction can be made subject to experimental manipulation.

Reduced metabolism in response to asphyxia can be demonstrated in reactions of animals ranging from less complex organisms to humans. Some mechanisms whereby these spontaneous reactions occur and are regulated have been subjects of study, but many details remain unclear. The decline in oxygen demand associated with a lowering of body temperature, roughly 10 percent per degree C, represents a considerable savings in metabolic activity and in the requirements for its support. Mortola (1993, 2004) describes this reaction in infant animals, comparing it with the reaction in adults of the same species, and demonstrates the resistance of the newborn animal to the consequences of asphyxia threat. His results support the evidence that protective reduction of metabolism at lowered body temperature is characteristic of a broad spectrum of species from protozoa to mammals (Wood 1991; Wood and Gonzales 1996).

Fetal and neonatal asphyxial tolerance

Tolerance of most mammals to the threat of arrested breathing, however modest when compared with what seals routinely experience, is especially noticeable in the mammalian fetus and newborn when they are threatened with disrupted gas exchange. The topic of mammalian fetal and neonatal reactions to asphyxia may appear to be remote from the discourse of this book that relates to the seal's diving and the yogi's meditating. However, the newborn mammal deserves inclusion among those that tolerate metabolic depression by virtue of its ability to endure this departure from what we consider equable life better than can older members of the same species. Its recourse to depressed metabolism makes the asphyxia-threatened newborn mammal a metabolic relative of the similarly responding submerged seal and meditator.

The near-term mammalian fetus must endure two challenges to its respiration that would be traumatic to the adult animal: intrauterine hypoxia and birth asphyxia (Singer 1999). Blood oxygen tension of the fetal circulation is lower than the maternal value, situated as it is at the low end of the respiratory cascade from atmosphere to fetus. Accordingly, arterial blood oxygen delivered to the fetus declines to values equivalent to those of high-altitude exposure. The resulting fetal reaction is one of adaptation to hypoxia as well as resistance to asphyxia.

Important aspects of fetal blood gas concentrations pertinent to mechanisms of resistance to hypoxia and asphyxia have been recognized, values existing during intrauterine development and changes relating to the abrupt alterations accompanying birth. Thus, fetal oxygen reserves are elevated by enhanced red blood cell concentrations, and thereby blood and tissue oxygen levels resemble those of high-altitude animals. But carbon dioxide and pH values tend toward those of the breath-holding diver, differing from the altitude dweller. They suggest a progression toward asphyxial levels rather than the relatively hypoxic–hypocapnic–alkalotic condition (low oxygen, low carbon dioxide, alkaline) resulting from the increased respiratory stimulus of altitude exposure. The distinction from the situation at altitude is worth noting, because the designation of hypoxia alone does not convey the entire condition. The combination of values continues during birth, further developing the triad asphyxial combination of hypoxia, hypercapnia, and acidosis before breathing is initiated and blood gas concentrations become more stabilized. Lung and blood gases change abruptly after the first newborn breath, and soon thereafter assume those of the independently living mammal.

The newborn animal must be prepared to react to the threat of asphyxia during birth, as it might be precipitated by the accompanying disruption of respiratory gas exchange through the placenta and umbilical cord before the newborn begins independent respiration. Several mechanisms contribute to protecting the continuity of its respiratory function. Increased blood oxygen stores in the form of elevated red cell concentration and a leftward shift of the hemoglobin dissociation curve provide the neonate with elevated blood oxygen capacity (Metcalfe, Moll, and Bartels 1964). The relatively undeveloped condition of the fetal brain, and thus its lesser metabolic demand, improves its tolerance of asphyxia. The neonatal heart also has better defenses against asphyxia and preserves tissue ATP better than does the adult heart. These responses have been determined from experimentally induced hypoxia in rabbit and dog neonates (Jarmakani et al. 1978a, 1978b), untested in human infants for obvious ethical reasons.

Protection of the newborn against asphyxia

Human protective responses to asphyxia, however modest when compared with what seals routinely experience, are nevertheless likely to play a role in life-sustaining defenses against those situations in which interrupted respiration poses a threat to survival. These protective mechanisms are especially noticeable when they are triggered into operation in the fetus and newborn in conditions that might result in disruption of respiratory gas exchange. The breath-hold diver is a more realistic model for the respiratory condition of the late-term mammalian fetus than the animal native to high altitude. The comparison is appropriate and reveals something of the contrasting mechanisms governing the transition to independent life after birth. In addition to the challenges of intrauterine life, the fetus must also be prepared to deal with the abrupt onset of asphyxia during birth and the accompanying disruption of respiratory gas exchange through the compressed placenta and umbilical cord.

Adjustments to the relatively low oxygen tension characteristic of uterine life and of birth asphyxia may be expected to protect the fetus and newborn animal against encounters with these hazards (Vannucci and Duffy 1974). Hypoxic preconditioning (discussed in chapter 6), the protective effect of prior exposure to moderate hypoxia on reducing subsequent ischemic damage to brain and other organs, has been demonstrated in neonatal rats (Gidday et al. 1994). Similar protection of the neonatal heart may result from induction of hypoxemia as a consequence of exposure by newborn rabbits to intermittent hypoxia from birth (Baker et al. 1995), and that capacity likely depends upon enhanced capability for utilization of cardiac muscle glycolysis by the neonatal myocardium (Dawes et al. 1959).

Reasons for the neonatal superiority in asphyxia tolerance, compared with adult vulnerability of the same species in reacting to this threat to metabolic stability, have been found in several lines of investigation. The newborn animal's primary response is one of reducing metabolism and the moment-to-moment demands of temperature regulation. These reactions suggest resemblances among neonates, hibernators, and diving mammals. Their defenses against threats to metabolic integrity similarly rely upon temporary suppression of the high immediate costs of active life (Bickler 2004). A first line of defense for the newborn mammal is a characteristic reduction of body temperature and consequent decline in metabolic rate. The similarity of this degree of imposed infant hypothermia and its resulting important protective effects to that which is experienced routinely by diving mammals is noteworthy and suggests a general principle of metabolic retreat in response to an impending threat.

The transition of the newborn from dependence on placental gas exchange to initiation of breathing with its own lungs may last for several minutes, leading inevitably toward respiratory disruption and progressive asphyxia. During this critical period the circulation is altered in a characteristic manner to protect the infant by preferential blood flow to the vital and oxygen-dependent central nervous system (Dawes 1968; Elsner, Hammond, and Parker 1970; Jensen and Berger 1991). An associated protective mechanism results from the general metabolic decline noticeable in newborn animals. The reaction to respiratory suspension during birth routinely amounts to a 30 percent reduction in oxygen consumption in newborn primates, the species most closely related to humans for which such information is available. There are obvious ethical reasons preventing related studies in human subjects, but such trials in primates suggest that a comparable life-saving metabolic suppression occurs in similarly exposed human infants.

The once-in-a-lifetime event of the mammal's birth is accompanied by circulatory reactions tending to protect the central nervous system and somewhat resembling those of the breath-hold diver. Characteristic protective cardiovascular reactions against the usual incidence of asphyxia are readily apparent. The immature condition of the neonatal central nervous system and its correspondingly reduced metabolic demand have been noted in several studies. The theme of their reactions is well summarized in a quote from Lutz, Nilsson, and Prentice (2003): "Mass specific cerebral oxygen and aerobic glucose consumption for the 7 day old rat is only one-tenth that of the adult and the rate of fall in brain ATP during anoxia is much less in the neonate."

When exposed to prolonged asphyxial conditions in which normal respiration is reduced, the newborn mammal responds to that condition by temporary reduction or elimination of that portion of its metabolism devoted to growth, a value estimated to be up to 40 percent of the total (Sidi et al. 1983). Further reaction involves its readiness for reducing or eliminating effective responses for thermoregulation, thus leading to the protective effects of body cooling and the consequently lowered metabolism (Mortola 1999, 2004). A resemblance to the hypoxia of high-altitude exposure is suggested. This condition of the newborn mammal differs somewhat from ordinary altitude exposure, because fetal blood carbon dioxide can be expected to be elevated above that resulting from the usual hyperventilation of altitude, and pH tends toward acidic rather than alkalotic conditions more resembling those experienced during long breath-holding. But the increase of carbon dioxide is at least partially offset by elimination due to increased breathing associated with an elevated level of circulating estrogen during pregnancy (Dejours 1981).

These conditions have been determined from experimentally induced hypoxia in rabbit and dog neonates (Jarmakani et al. 1978a, 1978b). The responses, resembling in some respects the reactions of both the diving mammal and the high-altitude dweller, are apparently essential for survival of the newly born animal when initiating lung inflation on its first breath. Despite the high fraction of body weight represented by the brain in these animals, the relatively undeveloped state of the newborn mammal's central nervous system is less metabolically active and therefore requires less oxygen consumption. The brains of one-week-old rats consume only a small fraction per gram of the adult brain oxygen requirement (Vannucci 1990), and there is accordingly less demand for ATP (Duffy et al. 1981).

If newborn mammals were to conform to the usual increase in metabolic rate of small animals, it would be anticipated that their metabolism would be accordingly elevated. However, the mature fetus does not show this increased metabolic rate, as dictated by small size, but rather acts like a part of the more slowly metabolizing maternal body. This condition is reversed shortly after birth, and it converts to the increased metabolic rate characteristic for the young animal (review: Dawes 1968). As a consequence of these temporary reactions that occur during the newborn animal's brief period of reduced metabolism before its postnatal increase, its tolerance of asphyxia can be temporarily enhanced and exceed that of the adult. However, the newborn's adaptive mechanisms against the threats of asphyxia are limited in scope, and the penalty for exceeding that capacity is revealed in the vulnerability of brain, and to a lesser degree kidney and intestine, to extended ischemia consequent to prolonged respiratory difficulty. Despite these problems, fetal and newborn mammals can, as indicated, take advantage of their temporarily reduced metabolic responses to the resulting modest body cooling by energy conservation (Mortola, Rezzonico, and Lanthier 1989; Bonora, Boule, and Gautier 1994; Thoresen and Wyatt 1997), an important reserve capacity in their defenses against asphyxia.

Hypothermia in the newborn

Purposeful lowering of body temperature has been sometimes successfully used in the treatment of metabolically depressed newborn human infants that failed to initiate breathing. Their unresponsive condition may be substantially improved to the extent that spontaneous respiration can be induced to begin after delays of several minutes; one successful trial was effective after a delay of seventy-nine minutes (Wyatt and Thoresen 2011)! Advantageous use of this technique is verified by the lack of subsequent impairment in a review of these patients ten years later. Protection of cellular integrity is a likely consequence of a lowered body temperature, resulting in reduced oxygen requirement as classically described by Krogh (1941). The survival effects of newborn hypothermia may not all be beneficial, and excessively forced intervention in lowering the infant's body temperature can have adverse cardiac and other effects. These complications may be avoided or reduced if blood pH is permitted to rise as temperature falls, in recognition of the temperature dependence of pH, increasingly alkaline with lowered temperature (Kroncke et al. 1986). Survival of the newborn animal might be enhanced by reduction of body temperature, resulting in reduced oxygen requirement. However, the survival enhancement of newborn hypothermia may not all be beneficial, and accidental or purposeful lowering of the infant's body temperature must be closely monitored.

The similarity of some characteristics of intermittent hypoxia and those of preconditioning is striking and suggests that they represent different aspects of the protection provided by repeated exposures to cycles of alternating free blood flow and circulatory restriction. Support for this view has been demonstrated in a study in which dogs were repeatedly exposed to brief periods of hypoxia equivalent to one-half sea-level atmospheric pressure but insufficiently long to produce hypoxic acclimatization responses, such as elevated blood hemoglobin and arterial oxygen content. The significant finding was that the intermittent hypoxia exposures were found to protect the heart muscle from infarction during subsequent experimental coronary artery occlusion (Zong et al. 2004).

Newborn lambs and seals

Resistance to hypoxia of the infant brain is consistent with the observation that it requires, perhaps unexpectedly, less oxygen and maintenance energy than the adult brain. A primary response is one of reducing metabolism and the moment-to-moment demands of temperature regulation. These reactions suggest resemblances among newborn, hibernating, and diving mammals. Their defenses against threats to metabolic integrity rely upon temporary suppression of the high immediate costs of active life (Bickler 2004). The ultimate limitations of the newborn's adaptive mechanisms against asphyxia are revealed in the vulnerability of brain, kidney, and intestine to extended ischemia that accompanies prolonged respiratory difficulty. Both the lamb and infant seal are protected against temporary asphyxial threat by preferential blood flow to the vital and oxygen-dependent central nervous system (Behrman et al. 1970; Elsner, Hammond, and Parker 1970; Liggins et al. 1980; Jensen and Berger 1991). A similar response is evident in human infants, a potentially life-saving metabolic retreat.

The neonatal dog brain tolerates oxygen lack better than its adult counterpart, as shown by its superior ability to endure longer total lack of blood flow (Kabat 1940). Adequate circulation is essential for long endurance of asphyxia in the newborn lamb, as survival depends on blood transport of glucose derived from cardiac muscle glycogen, thus providing substrate for anaerobic metabolism (Dawes et al. 1963). Brain tissue storage of glycogen, modest in most mammals, is unusually elevated in seals (Kerem, Hammond, and Elsner 1973). The advantage of this adequate supply when it is needed is illustrated by the enhanced hypoxic survival of rat brains when they are experimentally perfused with glucose (Vannucci and Mujsce 1992).

The newborn terrestrial mammal's tolerance of asphyxia demonstrates a general condition compared with adults of the same species. This characteristic is reversed in newborn seals and in the reactions of neonatal marine mammals generally. That is, their asphyxial tolerance does not markedly differ from that of the terrestrial newborn, but their diving endurance is much less than an adult of the same species can comfortably endure. This relatively modest tolerance of submersion in the newborn and juvenile seal has been verified in several species (Harrison and Tomlinson 1960; Irving et al. 1963; Hammond et al. 1969). Newborn Weddell seals, for example, routinely dive not longer than five minutes, far short of the adult capability (Kooyman 1968), appearing to resemble their neonatal terrestrial cousins in this respect. The newborn seal's modest diving endurance, compared with the adult capability, is likely accounted for by several anatomical and physiological aspects of its early life. Reactions of fetal and newborn mammals to asphyxia resulting in the slowing of heart rate have been noted for many years (Bauer 1937, 1938; Barcroft 1946). The blood of newborn seals lacks the adaptive high circulating hemoglobin level of the adult (Lenfant et al. 1969), the reverse of the situation in newborn land mammals. However, the maturing seal will eventually far exceed its terrestrial cousin in both blood volume and hemoglobin concentration as related to body weight.

Protective reactions

Practical applications of the described reactions in humans are important for real-life situations such as are encountered in the practice of emergency medicine. One such is the immediate treatment of severe trauma of the sort encountered in battlefield casualties and civilian accidents involving massive bleeding, cardiac arrest, and interrupted respiration. Immediate emergency responses to peripheral wounds of the extremities, even those involving dismemberment and life-threatening arterial hemorrhage, may in these situations sometimes be successfully managed by application of relatively simple techniques. But successful treatment of the more life-threatening internal and central nervous system traumas is, of course, much more likely to depend upon prompt intervention by skilled medical personnel.

In contrast, the mammalian brain reacts to oxygen deprivation by failure of osmotic volume adjustments dependent upon active regulatory mechanisms, and potentially damaging reactions ensue. Little margin for error exists before the driving force of arterial pressure may lead to brain ischemia resulting from constraint within the rigid skull. In these instances, the critical issue becomes the time between the moment of accident and adequate treatment. Successful recovery requires rapid reduction of intracranial pressure and surgical removal of accumulated blood.

A somewhat more radical suggestion has been proposed, that such a life-threatening condition may be amenable to successful outcome if cautious delay can be produced by metabolic suppression of the victim's conscious state. Such a procedure has been considered for facilitating recovery from severe trauma until more favorable options are available. Deep hypothermia, 10°C to 20° C, resulting in cardiac arrest and suspended animation, has been tested in animals as a short-term emergency method for rapidly reducing metabolism and preserving viability (Behringer et al. 2003; Roth and Nystul 2005; Wu et al. 2006). Immediate whole-body cooling and replacement of blood by a low-viscosity, nonclotting fluid might thus provide additional time for subsequent drastic interventions that could not otherwise be considered (Taylor et al. 1995).

Rapidly induced cooling, even to near-freezing levels, has been successfully used in experimental treatment of brain injury in dogs (Behringer et al. 2003). Similar protection from neurological damage of severely hypoxic human infants has been produced by body cooling (Cordey, Chiolero, and Miller 1973). Induction of controlled hypothermia in such emergency circumstances may be beneficial by virtue of its likely effects in addition to reducing oxygen requirement, such as conservation of ATP and lowered free radical generation with consequent neurological protection. It suggests a resemblance to the earlier related protective reaction that sustains viability in some instances of human survival during prolonged accidental immersion in cold water (Siebke et al. 1975; Sekar et al. 1980; Gilbert et al. 2000).

Beneficial depression of body temperature by a few degrees has been noted in some reports (Wood and Gonzales 1996; Walpoth and Daanen 2006). Forced deep hypothermia is generally hazardous, but it might be manageable by adjustment of acid-base balance during the cooling process (Kroncke et al. 1986). Moderate temperature reduction has been identified by Gerczuk and Kloner (2012) as one of the more effective techniques for the prompt treatment of myocardial infarction. It is noteworthy that some of the currently regarded relatively simple emergency treatments of this frequent cause of sudden death, conditioning and cooling, are easily induced, requiring modest training and relatively little equipment.

Some conclusions

Lessons learned from seal and human divers tell us something about the range and character of responses to be anticipated when the vastly differing lifestyles of these mammalian species engage in breath-hold submersion in water. The seal's more obvious cardiovascular changes while submerged support its metabolic adjustments that can influence the duration of the dive. Its longest dives usually entail little activity, the seal's endurance being dependent upon conservation of metabolic reserves, notably by reduced muscular exercise. A likely comparable human response is suggested to be the metabolic depression that occurs in deeply meditating yogis. Evidence for the seal's reactions is derived from extensive experimental results. What we know of the yogi's responses rests upon the results of only a few objective trials. Confidence in the reality and variations of the meditator's condition clearly will require a more substantial testing, but the prospects for useful results suggest that it may be an informative, perhaps even exciting, enterprise.

Research pertaining to the diving seal at rest and the meditator suggests that they share common physiological conditions that lead to metabolic conservation. My purpose has been to suggest that inquiry into reactions of these disparate species and conditions can lead to useful new understandings regarding the scope of responses to the particular challenges of their lives. Clearly, we are much further advanced in our understanding of the diving seal's reactions than we are in our perceptions of what happens in the meditating human. Much of our present knowledge of the diving seal's reactions has been acquired at an ever-increasing rate within recent decades, and many further questions have also been posed. Similar inquisitive attention to the human meditative condition is also likely to be a productive inquiry. It represents a rare, perhaps unique, instance of voluntary metabolic suppression by our species that approaches and may in some instances equal that of the quietly submerging seal.

The similarities and differences among the various experimental results described in these pages drawn from the examples of seals and yogis, and extended to hibernators and the newborn, raise questions with which biologists are familiar: Is one more impressed by the ways in which animals are alike or by differences among them? Is the category of lumping intellectually preferable to that of splitting? There are no easy answers to just how much liberty may be exercised in the attempt to identify hidden relationships and to generalize from their examples. Nevertheless, their judicious and sometimes speculative use can expose new, unheard-of, and exciting vistas—well stated by Jacob Bronowski (1964), who wrote that "all science is the search for unity in hidden likenesses."

REFERENCES

Anand, B. K., G. S. Chhina, and B. Singh. 1961. Studies on Shri Ramanand Yogi during his stay in an air-tight box. *Indian Journal of Medical Research* 48:82–89.

Andreka, G., M. Vertesaljai, G. Szantho, G. Font, Z. Piroth, G. Fontos, E. D. Juhasz, L. Szekely, Z. Szelid, M. S. Turner, H. Ashrafian, M. P. Frenneaux, and P. Andreka. 2007. Remote ischaemic postconditioning protects the heart during acute myocardial infarction in pigs. *Heart* 93:749–52.

Andersen, H. T. 1966. Physiological adaptations in diving vertebrates. *Physiological Reviews* 46:212–43.

Andrews, R. D., D. R. Jones, J. D. Williams, P. H. Thorson, G. W. Oliver, D. P. Costa, and B. J. Le Boeuf. 1997. Heart rates of northern elephant seals diving at sea and resting on the beach. *Journal of Experimental Biology* 200:2083–95.

Angell-James, J. E., R. Elsner, and M. de B. Daly. 1981. Lung inflation: Effects on heart rate, respiration, and vagal afferent activity in seals. *American Journal of Physiology* 240:H190–98.

Arnold, R. W. 1985. Extremes in human breath hold facial immersion bradycardia. *Undersea Biomedical Research* 12:183–90.

Asemu, G., F. Papousek, B. Ostadal, and F. Kolar. 1999. Adaptation to high altitude protects the rat heart against ischemia-induced arrhythmias: Involvement of mitochondrial K(ATP) channel. *Journal of Molecular and Cellular Cardiology* 31:1821–31.

Ashwell-Erickson, S., and R. Elsner. 1980. The energy cost of free existence for Bering Sea harbor and spotted seals. In *The Eastern Bering Sea Shelf: Oceanography and resources*, ed. D. W. Hood and J. A. Calder, 2:879–99. Seattle: University of Washington Press.

Austin, J. H. 1998. *Zen and the brain: Toward an understanding of meditation and consciousness.* Cambridge, MA: MIT Press.

———. 2006. *Zen-brain reflections: Reviewing recent developments in meditation and states of consciousness.* Cambridge, MA: MIT Press.

Baines, C. P., M. Goto, and J. M. Downey. 1997. Oxygen radicals released during ischemic preconditioning contribute to cardioprotection in the rabbit myocardium. *Journal of Molecular and Cellular Cardiology* 29:207–16.

Baker, E. J., L. E. Boerboom, G. N. Olinger, and J. E. Baker. 1995. Tolerance of the developing heart to ischemia: Impact of hypoxemia from birth. *American Journal of Physiology* 268:H1165–73.

Barcroft, J. 1946. *Researches on perinatal life*. Oxford: Blackwell.

Barnes, B. M. 1989. Freeze avoidance in a mammal: Body temperatures below 0°C in an Arctic hibernator. *Science* 244:1593–95.

Bauer, D. J. 1937. The slowing of heart rate produced by clamping the umbilical cord in the foetal sheep. *Journal of Physiology* 90:25P–27P.

———. 1938. The effect of asphyxia upon the heart rate of rabbits at different ages. *Journal of Physiology* 93:90–103.

Beall, C. M. 2007. Two routes to functional adaptation: Tibetan and Andean high-altitude natives. *Proceedings of the National Academy of Sciences of the USA* 104, suppl. 1: 8655–60.

Beary, J. F., and H. Benson. 1974. A simple psychophysiologic technique which elicits the hypometabolic changes of the relaxation response. *Psychosomatic Medicine* 36:115–20.

Béguin, P. C., M. Joyeux-Faure, D. Godin-Ribuot, P. Lévy, and C. Ribout. 2005. Acute intermittent hypoxia improves rat myocardium tolerance to ischemia. *Journal of Applied Physiology* 99:1064–69.

Behringer, W., P. Safar, X. Wu, R. Kenter, A. Radovsky, P. M. Kochanek, C. E. Dixon, and S. A. Tisherman. 2003. Survival without brain damage after clinical death of 60–120 mins in dogs using suspended animation by profound hypothermia. *Critical Care Medicine* 31:1523–31.

Behrman, R. E., M. H. Lees, E. N. Petersen, C. W. de Lannoy, and A. E. Seeds. 1970. Distribution of the circulation in the normal and asphyxiated fetal primate. *American Journal of Obstetrics and Gynecology* 108:956–69.

Benson, H. 1975. *The relaxation response*. New York: Avon Books.

Berger, R. J., and N. A. Phillips. 1988. Comparative aspects of energy metabolism, body temperature and sleep. *Acta Physiologica Scandinavica Supplementum* 574:21–27.

Bert, P. 1870. *Leçons sur le physiologie comparée de la respiration*. Paris: Bailliere et fils.

Bharma, S., and W. K. Milsom. 1993. Acidosis and metabolic rate in golden mantled ground squirrels (*Spermophilus lateralis*). *Respiration Physiology* 94:337–51.

Bickler, P. E. 1984. CO_2 balance of a heterothermic rodent: Comparison of sleep, torpor, and awake states. *American Journal of Physiology* 246:R49–55.

———. 2004. Clinical perspectives: Neuroprotection lessons from hypoxia-tolerant organisms. *Journal of Experimental Biology* 207:3243–49.

Bing, O. H. L., W. W. Brooks, and J. V. Messer. 1973. Heart muscle viability following hypoxia: Protective effect of acidosis. *Science* 180:1297–98.

Birnbaum, Y., S. L. Hale, and R. A. Kloner. 1997. Ischemic preconditioning at a distance: Reduction of myocardial infarct size by partial reduction of blood supply combined with rapid stimulation of the gastrocnemius muscle in the rabbit. *Circulation* 96:1641–46.

Blackstone, E., M. Morrison, and M. B. Roth. 2005. H_2S induces a suspended animation-like state in mice. *Science* 308:518.

Blix, A. S., and B. Folkow. 1983. Cardiovascular adjustments to diving in mammals and birds. In *Handbook of physiology*, sec. 2, *The cardiovascular system*, ed. T. Shepherd and F. M. Abboud, 917–45. Bethesda, MD: American Physiological Society.

Bolli, R. 2007. Preconditioning: A paradigm shift in the biology of myocardial ischemia. *American Journal of Physiology* 292:H19–27.

Bolling, S. F., N. L. Tramontini, K. S. Kilgore, T-P. Su, P. R. Oeltgen, and H. H. Harlow. 1997. Use of "natural" hibernation induction triggers for myocardial protection. *Annals of Thoracic Surgery* 64:623–27.

Bolte, R. G., P. G. Black, R. S. Bowers, J. K. Thorne, and H. M. Cornell. 1988. The use of extracorporeal rewarming in a child submerged for 66 minutes. *Journal of the American Medical Association* 260:377–79.

Bonventre, J. V. 2002. Kidney ischemic preconditioning. *Current Opinion in Nephrology and Hypertension* 11:43–48.

Bonora, M., M. Boule, and H. Gautier. 1994. Ventilatory strategy in hypoxic or hypercapnic newborns. *Biology of the Neonate* 65:198–204.

Boorstin, D. 1983. *The discoverers.* New York: Random House.

Boutilier, R. G. 2001. Mechanisms of metabolic defense against hypoxia in hibernating frogs. *Respiration Physiology* 128:365–77.

Boyer, B. B., and B. M. Barnes. 1999. Molecular and metabolic aspects of mammalian hibernation. *Bioscience* 49:713–24.

Boyle, R. 1670. New pneumatical experiments about respiration. *Philosophical Transactions of the Royal Society* 5:2011–31.

Brinkmann, L., M. Gerken, and A. Riek. 2012. Adaptation strategies to seasonal changes in environmental conditions of a domesticated horse breed, the Shetland pony (*Equus ferus caballus*). *Journal of Experimental Biology* 215:1061–68.

Broad, W. J. 2012. *The science of yoga: The risks and the rewards.* New York: Simon and Schuster.

Brodie, P., and A. Paasche. 1985. Thermoregulation and energetics of fin and sei whales based on postmortem, stratified temperature measurements. *Canadian Journal of Zoology* 63:2267–69.

Bronowski, J. 1964. *Science and human values.* London: Pelican.

Bunn, H. F., and R. O. Poyton. 1996. Oxygen sensing and molecular adaptation to hypoxia. *Physiological Reviews* 76:839–85.

Burmester, T., and T. Hankeln. 2004. Neuroglobin: A respiratory protein of the nervous system. *News in Physiological Sciences* 19:110–13.

Burns, J. M., K. C. Lestyk, L. P. Folkow, M. O. Hammill, and A. S. Blix. 2007. Size and distribution of oxygen stores in harp and hooded seals from birth to maturity. *Journal of Comparative Physiology B* 177:687–700.

Burns, J. M., N. Skomp, N. Bishop, K. Lestyk, and M. Hammill. 2010. Development of aerobic and anaerobic metabolism in cardiac and skeletal muscles from harp and hooded seals. *Journal of Experimental Biology* 213:740–48.

Bushell, W. C., E. L. Olivo, and N. D. Theise, eds. 2009. *Longevity, regeneration and optimal health: Integrating Eastern and Western perspectives.* Vol. 1172 of *Annals of the New York Academy of Sciences.*

Butler, P. J. 1990. Metabolic adjustments to breath holding in higher vertebrates. *Canadian Journal of Zoology* 67:3024–31.

———. 2004. Metabolic regulation in diving birds and mammals. *Respiration Physiology and Neurobiology* 141:297–315.

Butler, P. J., and D. R. Jones. 1982. The comparative physiology of diving in vertebrates. *Advances in Comparative Physiology and Biochemistry* 8:179–364.

———. 1997. Physiology of diving of birds and mammals. *Physiological Reviews* 77:837–99.

Cai, Z., D. J. Manalo, G. Wei, E. R. Rodriguez, K. Fox-Talbot, H. Lu, J. L. Zweier, and G. L. Semenza. 2003. Hearts from rodents exposed to intermittent hypoxia or erythropoietin are protected against ischemia-reperfusion injury. *Circulation* 108:79–85.

Carey, H. V., M. T. Andrews, and S. L. Martin. 2003. Mammalian hibernation: Cellular and molecular responses to depressed metabolism and low temperature. *Physiological Reviews* 83:1153–81.

Castellini, J. M., and M. A. Castellini. 2012. Life under water: Physiological adaptations to diving and living at sea. *Comprehensive Physiology* 2:1889–1919.

Castellini, M. A. 1985. Metabolic depression in tissues and organs of marine mammals during diving: Living longer with less oxygen. *Molecular Physiology* 8:427–37.

———. 1996. Dreaming about diving: Sleep apnea in seals. *News in Physiological Sciences* 11:208–14.

Castellini, M. A., and J. M. Castellini. 2004. Defining the limits of diving biochemistry in marine mammals. *Comparative Biochemistry and Physiology B* 139:509–18.

Castellini, M. A., R. W. Davis, and G. L. Kooyman. 1988. Blood chemistry regulation during repetitive diving in Weddell seals. *Physiological Zoology* 61:379–86.

Castellini, M. A., G. L. Kooyman, and P. J. Ponganis. 1992. Metabolic rates of freely diving Weddell seals: Correlations with oxygen stores, swim velocity and diving duration. *Journal of Experimental Biology* 165:181–94.

Castellini, M. A., B. J. Murphy, M. Fedak, K. Ronald, N. Gofton, and P. W. Hochachka. 1985. Potentially conflicting metabolic demands of diving and exercise in seals. *Journal of Applied Physiology* 251:392–99.

Castellini, M. A., and G. N. Somero. 1981. Buffering capacity of vertebrate muscle: Correlations with potentials for anaerobic function. *Journal of Comparative Physiology* 143:191–98.

Chien, G. L., R. A. Wolff, R. F. Davis, and D. M. Van Winkle. 1994. "Normothermic range" temperature affects myocardial infarct size. *Cardiovascular Research* 28:1014–17.

Cohen, M. V., C. P. Baines, and J. M. Downey. 2000. Ischemic preconditioning: From adenosine receptor to KATP channel. *Annual Review of Physiology* 62:79–109.

Cohen, M. V., and J. M. Downey. 2008. Adenosine: Trigger and mediator of cardioprotection. *Basic Research in Cardiology* 103:203–15.

Cohen, M. V., X. M. Yang, and J. M. Downey. 2007. The pH hypothesis of postconditioning: Staccato reperfusion reintroduces oxygen and perpetuates myocardial acidosis. *Circulation* 115:1895–903.

Cordey, R., R. Chiolero, and J. A. Miller. 1973. Resuscitation of neonates by hypothermia: Report on 20 cases with acid-base determination on 10 cases and the long term development of 33 cases. *Resuscitation* 2:169–81.

Crile, G., and D. P. Quiring. 1940. A record of the body weight and certain organ and gland weights of 3690 animals. *Ohio Journal of Science* 40:219–59.

Cross, E. R. 1965. Taravana—diving syndrome in the Tuamotu diver. In Rahn 1965, 207–19.

Currin, R. T., G. J. Gores, R. G. Thurman, and J. J. Lemasters. 1991. Protection by acidotic pH against anoxic cell killing in perfused rat liver: Evidence for a pH paradox. *FASEB Journal* 5:207–10.

Daan, S., B. M. Barnes, and A. M. Strijkstra. 1991. Warming up for sleep? Ground squirrels sleep during arousals from hibernation. *Neuroscience Letters* 128:265–68.

Daly, M. de B. 1984. Breath-hold diving: Mechanisms of cardiovascular adjustments in the mammal. In *Recent advances in physiology*, ed. P. F. Baker, 201–45. London: Churchill Livingstone .

———. 1997. *Peripheral arterial chemoreceptors and respiratory-cardiovascular integration*. Monographs of the Physiological Society no. 46. Oxford: Clarendon Press.

Daly, M. de B., and J. E. Angell-James. 1975. Role of the arterial chemoreceptors in the control of the cardiovascular responses to breath-hold diving. In *The peripheral arterial chemoreceptors*, ed. M. J. Purves, 387-407. London: Cambridge University Press.

Daly, M. de B., R. Elsner, and J. E. Angell-James. 1977. Cardiorespiratory control by carotid chemoreceptors during experimental dives in the seal. *American Journal of Physiology* 232:H508–16.

Dave, K. R., R. Prado, A. P. Raval, K. L. Drew, and M. A. Pérez-Pinzón. 2006. The arctic ground squirrel brain is resistant to injury from cardiac arrest during euthermia. *Stroke* 37:1261–65.

Davidson, J. M. 1976. The physiology of meditation and mystical states of consciousness. *Perspectives in Biology and Medicine* 19:345–79.

Davies, D. G. 1989. Distribution of systemic blood flow during anoxia in the turtle, *Chrysemys scripta*. *Respiration Physiology* 78:383–90.

Davies, P. 1999. *The fifth miracle: The search for the origin and meaning of life*. New York: Simon and Schuster.

Davis, R. W. 2014. A review of the multi-level adaptations for maximizing aerobic dive duration in marine mammals: From biochemistry to behavior. *Journal of Comparative Physiology B* 184:23–53.

Davis, R. W., L. Polasek, R. Watson, A. Fuson, T. M. Williams, and S. B. Kanatous. 2004. The diving paradox: New insights into the role of the dive response in air-breathing vertebrates. *Comparative Biochemistry and Physiology A* 138:263–68.

Davis, R. W., and T. M. Williams. 2012. The marine mammal dive response is exercise modulated to maximize aerobic dive duration. *Journal of Comparative Physiology A* 198:583–91.

Dawes, G. S. 1968. *Foetal and neonatal physiology*. Chicago: Year Book Medical Publishers.

Dawes, G. S., H. N. Jacobson, J. C. Mott, J. H. Shelley, and A. Stafford. 1963. The treatment of asphyxiated, mature foetal lambs and rhesus monkeys with intravenous glucose and sodium carbonate. *Journal of Physiology* 169:167–84.

Dawes, G. S., J. C. Mott, and H. J. Shelley. 1959. The importance of cardiac glycogen for the maintenance of life in foetal lambs and new-born animals during anoxia. *Journal of Physiology* 146:516–38.

Dejours, P. 1981. *Principles of Comparative Respiratory Physiology.* 2nd ed. Amsterdam: Elsevier/North-Holland.

Donohoe, P. H., and R. G. Boutilier. 1998. The protective effects of metabolic rate depression in hypoxic cold submerged frogs. *Respiration Physiology* 111:325–36.

Douglas, R. M., and G. G. Haddad. 2008. Can O_2 dysregulation induce premature aging? *Physiology* 23:333–49.

Drew, K. L., C. L. Buck, B. M. Barnes, S. L. Christian, B. T. Rasley, and M. B. Harris. 2007. Central nervous system regulation of mammalian hibernation: Implications for metabolic suppression and ischemia tolerance. *Journal of Neurochemistry* 102:1713–26.

Drew, K. L., M. E. Rice, T. B. Kuhn, and M. A. Smith. 2001. Neuroprotective adaptations in hibernation: Therapeutic implications for ischemia-reperfusion, traumatic brain injury and neurodegenerative diseases. *Free Radical Biology and Medicine* 31:563–73.

Duffy, T. E., M. Cavazutti, N. F. Cruz, and L. Sokoloff. 1981. Local cerebral glucose metabolism in newborn dogs: Effects of hypoxia and halothane anesthesia. *Annals of Neurology* 11:233–46.

Elsner, R. 1963. Comparison of Australian Aborigines, Alakaluf Indians and Andean Indians. In *Proceedings of the International Symposium on Temperature Acclimation. Federation Proceedings* 22:840–42.

———. 1965. Heart rate responses in forced versus trained experimental dives in pinnipeds. *Hvalrådets Skrifter* 48:24–29.

———. 1986. Limits to exercise performance: Some ideas from comparative studies. *Acta Physiologica Scandinavica Supplementum* 556:44–51.

———. 1999. Living in water: Solutions to physiological problems. In *Biology of Marine Mammals*, ed. J. E. Reynolds III and S. A. Rommel, 73–116. Washington, DC: Smithsonian Institution.

Elsner, R., and M. de B. Daly. 1988. Coping with asphyxia: Lessons from seals. *News in Physiological Sciences* 3:65–69.

Elsner, R., D. L. Franklin, and R. L. Van Citters. 1964. Cardiac output during diving in an unrestrained sea lion. *Nature* 202:809–10.

Elsner, R., D. L. Franklin, R. L. Van Citters, and D. W. Kenney. 1966. Cardiovascular defense against asphyxia. *Science* 153:941–49.

Elsner, R., and B. Gooden. 1983. *Diving and asphyxia: A comparative study of animals and man.* Monographs of the Physiological Society no. 40. Cambridge: Cambridge University Press.

Elsner, R., D. D. Hammond, and H. R. Parker. 1970. Circulatory responses to asphyxia in pregnant and fetal animals: A comparative study of Weddell seals and sheep. *Yale Journal of Biology and Medicine* 42:202–17.

Elsner, R., D. W. Kenney, and K. Burgess. 1966. Diving bradycardia in the trained dolphin. *Nature* 212:407–8.

Elsner, R., R. W. Millard, J. K. Kjekshus, F. White, A. S. Blix, and W. S. Kemper. 1985. Coronary blood flow and myocardial segment dimensions during simulated dives in seals. *American Journal of Physiology* 249:H1119–26.

Elsner, R., S. Øyasæter, R. Almaas, and O. D. Saugstad. 1998. Diving seals, ischemia-reperfusion and oxygen radicals. *Comparative Biochemistry and Physiology A* 119:1019–25.

Elsner, R., J. T. Shurley, D. D. Hammond, and R. E. Brooks. 1970. Cerebral tolerance to hypoxemia in asphyxiated Weddell seals. *Respiration Physiology* 9:287–97.

Elsner, R., D. Wartzok, N. B. Sonafrank, and B. P. Kelly. 1989. Behavioral and physiological reactions of arctic seals during under-ice pilotage. *Canadian Journal of Zoology* 67:2506–13.

Epel, E., J. Daubenmier, J. T. Moskowitz, S. Folkman, and E. Blackburn. 2009. Can meditation slow rate of cellular aging? Cognitive stress, mindfulness, and telomeres. *Annals of the New York Academy of Sciences* 1172:34–53.

Fahey, J. T., and G. Lister. 1989. Response to low cardiac output: Developmental differences in metabolism during oxygen deficit and recovery in lambs. *Pediatric Research* 26:115–24.

Falke, K. J., R. D. Hill, J. Qvist, R. C. Schneider, M. Guppy, G. C. Liggins, P. W. Hochachka, R. E. Elliott, and W. M. Zapol. 1985. Seal lungs collapse during free diving: Evidence from arterial nitrogen tensions. *Science* 229:556–58.

Farrow, J. T., and J. R. Hebert. 1982. Breath suspension during the transcendental meditation technique. *Psychosomatic Medicine* 44:133–53.

Fedak, M. 2009. In situ blood oxygen analysis for a truly free-living deep-diving animal: Focus on "Extreme hypoxemic tolerance and blood oxygen depletion in diving elephant seals." *American Journal of Physiology* 297:R925–26.

Fedak, M. A., M. R. Pullen, and J. Kanwisher. 1988. Circulatory responses of seals to periodic breathing: Heart rate and breathing during exercise and diving in the laboratory and open sea. *Canadian Journal of Zoology* 66:53–60.

Fisher, D. J., M. A. Heymann, and A. M. Rudolph. 1982. Fetal myocardial oxygen and carbohydrate consumption during acutely induced hypoxemia. *American Journal of Physiology* 242:H657–61.

Florant, G. L., B. M. Turner, and H. C. Heller. 1978. Temperature regulation during wakefulness, sleep and hibernation in marmots. *American Journal of Physiology* 235: R82–88.

Folkow, L. P., J. M. Ramirez, S. Ludvigsen, N. Ramirez, and A. S. Blix. 2008. Remarkable neural hypoxia tolerance in the deep-diving adult hooded seal (*Cystophora cristata*). *Neuroscience Letters* 446:147–50.

Forman, M. B., C. E. Velasco, and E. K. Jackson. 1993. Adenosine attenuates reperfusion injury following regional myocardial ischaemia. *Cardiovascular Research* 27:9–17.

Frappell, P., C. Saiki, and J. P. Mortola. 1991. Metabolism during normoxia, hypoxia and recovery in the newborn kitten. *Respiration Physiology* 86:115–24.

Frappell, P., C. Lanthier, R. V. Baudinette, and J. P. Mortola. 1992. Metabolism and ventilation in acute hypoxia: A comparative analysis in small mammalian species. *American Journal of Physiology* 262:R1040–46.

Gamperl, A. K., A. E. Todgham, W. S. Parkhouse, R. Dill, and A. P. Farrell. 2001. Recovery of trout myocardial function following anoxia: Preconditioning in a non-mammalian model. *American Journal of Physiology* 281:R1755–63.

Garruto, R. M., and J. S. Dutt. 1983. Lack of prominent compensatory polycythemia in traditional native Andeans living at 4,200 meters. *American Journal of Physical Anthropology* 61:355–66.

Geiser, F. 2004. Metabolic rate and body temperature reduction during hibernation and daily torpor. *Annual Review of Physiology* 66:239–74.

Gerczuk, P. Z., and R. A. Kloner. 2012. An update on cardioprotection: A review of the latest adjunctive therapies to limit myocardial infarction size in clinical trials. *Journal of the American College of Cardiology* 59:969–78.

Gho, B. C. G., R. G. Schoemaker, M. A. van den Doel, D. J. Duncker, and P. D. Verdouw. 1996. Myocardial protection by brief ischemia in noncardiac tissue. *Circulation* 94:2193–200.

Gidday, J. M., J. C. Fitzgibbons, A. R. Shah, and T. S. Park. 1994. Neuroprotection from ischemic brain injury by hypoxic preconditioning in the neonatal rat. *Neuroscience Letters* 168:221–24.

Gilbert, M., R. Busund, A. Skagseth, P. Å. Nilsen, and J. P. Solbø. 2000. Resuscitation from accidental hypothermia of 13.7°C with circulatory arrest. *Lancet* 335:375–76.

Gooden, B. A. 1992. Why some people do not drown: Hypothermia versus the diving response. *Medical Journal of Australia* 157:629–32.

Granfeldt, A., D. J. Lefer, and J. Vinten-Johansen. 2009. Protective ischaemia in patients: Preconditioning and postconditioning. *Cardiovascular Research* 83:234–46.

Guppy, M., R. D. Hill, R. C. Schneider, J. Qvist, G. C. Liggins, W. M. Zapol, and P. W. Hochachka. 1986. Microcomputer-assisted metabolic studies of voluntary diving of Weddell seals. *American Journal of Physiology* 250:R175–87.

Guzy, R. D., and P. T. Schumacker. 2006. Oxygen sensing by mitochondria at complex III: The paradox of increased reactive oxygen species during hypoxia. *Experimental Physiology* 91:807–19.

Halasz, N. A., R. Elsner, R. S. Garvie, and G. T. Grotke. 1974. Renal recovery from ischema: A comparative study of harbor seal and dog kidneys. *American Journal of Physiology* 227:1331–35.

Hammel, H. T. 1964. Terrestrial animals in cold: Recent studies of primitive man. In *Handbook of physiology*, sec. 4, *Adaptation to the environment*, ed. D. B. Dill, 413–33. Washington, DC: American Physiological Society.

Hammel, H. T., R. Elsner, H. C. Heller, J. A. Maggert, and C. R. Bainton. 1977. Thermoregulatory responses to altering hypothalamic temperature in the harbor seal. *American Journal of Physiology* 232:R18–26.

Hammel, H. T., R. W. Elsner, D. H. Le Messurier, H. T. Andersen, and F. A. Milan. 1959. Thermal and metabolic responses of the Australian aborigine exposed to moderate cold in summer. *Journal of Applied Physiology* 14:605–15.

Hammond, D. D., R. Elsner, G. Simison, and R. Hubbard. 1969. Submersion bradycardia in the newborn elephant seal *Mirounga angustirostris*. *American Journal of Physiology* 216:220–23.

Harrison, R. J., and J. D. W. Tomlinson. 1960. Normal and experimental diving in the common seal *Phoca vitulina*. *Mammalia* 24:386–99.

Harvey, W. 1653. *An anatomical disputation concerning the movement of the heart and blood in living creatures*. Trans. from Latin by Gweneth Whitteridge. Oxford and London: Blackwell Scientific Publications, 1976.

Hausenloy, D. J., and D. M. Yellon. 2007. The evolving story of "conditioning" to protect against acute myocardial ischaemia-reperfusion injury. *Heart* 93:649–51.

———. 2008a. Preconditioning and postconditioning: New strategies for cardioprotection. *Diabetes, Obesity and Metabolism* 10:451–59.

———. 2008b. Remote ischaemic preconditioning: Underlying mechanisms and clinical application. *Cardiovascular Research* 79:377–86.

Heath, D., and D. R. Williams. 1981. *Man at high altitude: Pathophysiology of acclimatization and adaptation.* Edinburgh: Churchill Livingstone.

Heinrich, B. 2003. *Winter world: The ingenuity of animal survival.* New York: HarperCollins.

Heldmaier, G., S. Ortmann, and R. Elvert. 2004. Natural hypometabolism during hibernation and daily torpor in mammals. *Respiration Physiology and Neurobiology* 141:317–29.

Heller, H. C. 1988. Sleep and hypometabolism. *Canadian Journal of Zoology* 66:61–69.

Heller, H. C., and G. W. Colliver 1974. CNS regulation of body temperature during hibernation. *American Journal of Physiology* 227:583–89.

Heller, H. C., R. Elsner, and N. Rao. 1987. Voluntary hypometabolism in an Indian yogi. *Journal of Thermal Biology* 12:171–73.

Heller, H. C., G. L. Florant, S. F. Glotzbach, J. M. Walker, and R. J. Berger. 1978. Sleep and torpor—homologous adaptations for energy conservation. In *Dormancy and developmental arrest,* ed. M. E. Clutter, 269–96. New York: Academic Press

Heller, H. C., and S. F. Glotzbach. 1977. Thermoregulation during sleep and hibernation. *International Review of Physiology* 15:147–88.

Heller, H. C., and N. F. Ruby. 2004. Sleep and circadian rhythms in mammalian torpor. *Annual Review of Physiology* 66:275–89.

Hemmingsen, E. 1963. Enhancement of oxygen transport by myoglobin. *Comparative Biochemistry and Physiology* 10:239–44.

Hermes-Lima, M., and K. B. Storey. 1993. Antioxidant defenses in the tolerance of freezing and anoxia by garter snakes. *American Journal of Physiology* 265:R646–52.

Hermes-Lima, M., J. M. Storey, and K. B. Storey. 1998. Antioxidant defenses and metabolic depression: The hypothesis of preparation for oxidative stress in land snails. *Comparative Biochemistry and Physiology B* 120:437–48.

Hermes-Lima, M., and T. Zenteno-Savín. 2002. Animal response to drastic changes in oxygen availability and physiological oxidative stress. *Comparative Biochemistry and Physiology C* 133:537–56.

Hill, R. D., R. C. Schneider, G. C. Liggins, A. H. Schuette, R. L. Elliott, M. Guppy, P. W. Hochachka, J. Qvist, K. J. Falke, and W. M. Zapol. 1987. Heart rate and body temperature during free diving of Weddell seals. *American Journal of Physiology* 253:R344–51.

Hindell, M. A., D. J. Slip, and H. R. Burton. 1991. The diving behaviour of adult male and female southern elephant seals, *Mirounga leonina. Australian Journal of Zoology* 39: 595–619.

Hindell, M. A., D. J. Slip, H. R. Burton, and M. M. Bryden. 1992. Physiological implications of continuous, prolonged and deep dives of the southern elephant seal (*Mirounga leonina*). *Canadian Journal of Zoology* 70:370–79.

Hochachka, P. W. 1986. Balancing conflicting metabolic demands of exercise and diving. *Federation Proceedings* 45:2948–52.

Hochachka, P. W., and M. Guppy. 1987. *Metabolic arrest and the control of biological time.* Cambridge, MA: Harvard University Press.

Hochachka, P. W., and P. L. Lutz. 2001. Mechanism, origin, and evolution of anoxia tolerance in animals. *Comparative Biochemistry and Physiology B* 130:435–59.

Hochachka, P. W., J. L. Rupert, and C. Monge. 1999. Adaptation and conservation of physiological systems in the evolution of human hypoxia tolerance. *Comparative Biochemistry and Physiology A* 120:1–17.

Hoit, B. D., N. D. Dalton, S. C. Erzurum, D. Laskowski, K. P. Strohl, and C. M. Beall. 2005. Nitric oxide and cardiopulmonary hemodynamics in Tibetan highlanders. *Journal of Applied Physiology* 99:1796–1801.

Hong, S. K. 1963. Comparison of diving and non-diving women of Korea. *Federation Proceedings* 22:831–33.

———. 1973. Pattern of cold adaptation in women divers of Korea (ama). *Federation Proceedings* 32:1614–22.

———. 1989. Mechanism of tolerance to renal ischemia in harbor seal: Role of membranes. *Undersea Biomedical Research* 16:381–90.

Hong, S. K., S. Ashwell-Erickson, P. Gigliotti, and R. Elsner. 1982. Effects of anoxia and low pH on organic ion transport and electrolyte distribution in harbor seal (*Phoca vitulina*) kidney slices. *Journal of Comparative Physiology* 149:19–24.

Hong, S. K., D. W. Rennie, and Y. S. Park. 1986. Cold acclimatization and deacclimatization of Korean women divers. *Exercise and Sport Sciences Reviews* 14:231–68.

Hong, S. K., and H. Rahn. 1967. The diving women of Korea and Japan. *Scientific American* 216 (5): 34–43.

Hori, M., M. Kitakaze, H. Sato, S. Takashima, K. Iwakura, M. Inoue, A. Kitabatake, and T. Kamada. 1991. Staged reperfusion attenuates myocardial stunning in dogs: Role of transient acidosis during early reperfusion. *Circulation* 84:2135–45.

Hurford, W. E., P. W. Hochachka, R. C. Schneider, G. P. Guyton, K. S. Stanek, D. G. Zapol, G. C. Liggins, and W. M. Zapol. 1996. Splenic contraction, catecholamine release, and blood volume redistribution during diving in the Weddell seal. *Journal of Applied Physiology* 80:298–306.

Hurley, J. A., and D. P. Costa. 2001. Standard metabolic rate at the surface and during trained submersions in adult California sea lions (*Zalophus californianus*). *Journal of Experimental Biology* 204:3273–81.

Husted, T. L., W. Chang, A. B. Lentsch, and S. M. Rudich. 2004. δ-Opiod agonists protect the rat liver from cold storage and ischemia/reperfusion injury. In *Life in the cold: Evolution, mechanisms, adaptation, and application*, ed. B. M. Barnes and H. V. Carey, 575–84. 12th International Hibernation Symposium. Biological Papers, University of Alaska, no. 27. Fairbanks: Institute of Arctic Biology.

Irving, L. 1939. Respiration in diving mammals. *Physiological Reviews* 19:112–34.

Irving, L., P. F. Scholander, and S. W. Grinnell. 1942. The regulation of blood pressure in the seal during diving. *American Journal of Physiology* 135:557–66.

Irving, L., L. J. Peyton, C. H. Bahn, and R. S. Peterson. 1963. Action of the heart and breathing during development of fur seals (*Callorhinus ursinus*). *Physiological Zoology* 36:1–20.

Jackson, D. C. 2000. Living without oxygen: Lessons from the freshwater turtle. *Comparative Biochemistry and Physiology A* 125:299–315.

———. 2004. Overwintering in submerged turtles. In *Life in the cold: Evolution, mech-*

anisms, adaptation, and application, ed. B. M. Barnes and H. V. Carey, 317–27. 12th International Hibernation Symposium. Biological Papers, University of Alaska, no. 27. Fairbanks: Institute of Arctic Biology.

Jarmakani, J. M., M. Nakazawa, T. Nagatomo, and G. A. Langer. 1978a. Effect of hypoxia on mechanical function in the mammalian neonatal heart. *American Journal of Physiology* 235:H469–74.

———. 1978b. Effects of hypoxia on myocardial high-energy phosphates in the neonatal mammalian heart. *American Journal of Physiology* 235:H475–81.

Jennings, R. B., L. Sebbag, L. M. Schwartz, M. S. Crago, and K. A. Reimer. 2001. Metabolism of preconditioned myocardium: Effect of loss and reinstatement of cardioprotection. *Journal of Molecular and Cellular Cardiology* 33:1571–88.

Jensen, A., and R. Berger. 1991. Fetal circulatory responses to oxygen lack. *Journal of Developmental Physiology* 16:181–207.

Jevning, R., R. K. Wallace, and M. Beidebach. 1992. The physiology of meditation—a review: A wakeful hypometabolic integrated response. *Neuroscience and Biobehavioral Reviews* 16:415–24.

Jobsis, P. D., P. J. Ponganis, and G. L. Kooyman. 2001. Effects of training on forced submersion responses in harbor seals. *Journal of Experimental Biology* 204:3877–85.

Johansson, B. W. 1996. The hibernator heart—nature's model of resistance to ventricular fibrillation. *Cardiovascular Research* 31:826–32.

Johnson, P., R. Elsner, and T. Zenteno-Savín. 2004. Hypoxia-inducible factor in ringed seal (*Phoca hispida*) tissues. *Free Radical Research* 38:847–54.

———. 2005. Hypoxia-inducible factor-1 proteomics and diving adaptations in ringed seal. *Free Radical Biology and Medicine* 39:205–12.

Jones, D. R. 1976. The control of breathing in birds with particular reference to the initiation and maintenance of diving apnea. *Federation Proceedings* 35:1975–82.

Kabat, H. 1940. The greater resistance of very young animals to arrest of the brain circulation. *American Journal of Physiology* 130:588–99.

Kagen, L. J., and C. L. Christian. 1966. Immunologic measurements of myoglobin in human adult and fetal skeletal muscle. *American Journal of Physiology* 211:656–60.

Kanatous, S. B., L. V. DiMichele, D. F. Cowan, and R. W. Davis. 1999. High aerobic capacities in the skeletal muscles of pinnipeds: Adaptations to diving hypoxia. *Journal of Applied Physiology* 86:1247–56.

Kanatous, S. B., R. Elsner, and O. Mathieu-Costello. 2001. Muscle capillary supply in harbor seals. *Journal of Applied Physiology* 90:1919–26.

Kanatous, S. B., T. J. Hawke, S. J. Trumble, L. E. Pearson, R. R. Watson, D. J. Garry, T. M. Williams, and R. W. Davis. 2008. The ontogeny of aerobic and diving capacity in the skeletal muscles of Weddell seals. *Journal of Experimental Biology* 211:2559–65.

Kang, B. S., S. H. Song, C. S. Suh, and S. K. Hong. 1963. Changes in body temperature and basal metabolic rate of the Ama. *Journal of Applied Physiology* 18:483–88.

Karimi, A., K. T. Ball, and G. G. Power. 1996. Exogenous infusion of adenosine depresses whole body O_2 use in fetal/neonatal sheep. *Journal of Applied Physiology* 81:541–47.

Kerem, D., and R. Elsner. 1973a. Cerebral tolerance to asphyxial hypoxia in the harbor seal. *Respiration Physiology* 19:188–200.

————. 1973b. Cerebral tolerance to asphyxial hypoxia in the dog. *American Journal of Physiology* 225:593–600.

Kerem, D., D. D. Hammond, and R. Elsner. 1973. Tissue glycogen levels in the Weddell seal, *Leptonychotes weddelli*: A possible adaptation to asphyxial hypoxia. *Comparative Biochemistry and Physiology A* 45:731–36.

Kerendi, F., H. Kin, M. E. Halkos, R. Jiang, A. J. Zatta, Z. Q. Zhao, R. A. Guyton and J. Vinten-Johansen. 2005. Remote postconditioning. *Basic Research in Cardiology* 100: 404–12.

Kesterson, J., and N. F. Clinch. 1989. Metabolic rate, respiratory exchange ratio, and apneas during meditation. *American Journal of Physiology* 256:R632–38.

Kietzmann, T., J. Fandrey, and H. Acker. 2000. Oxygen radicals as messengers in oxygen-dependent gene expression. *News in Physiological Sciences* 15:202–8.

Kilduff, T. S., B. Krilowicz, W. K. Milsom, L. Trachsel, and L. C. Wang. 1993. Sleep and mammalian hibernation: Homologous adaptations and homologous processes? *Sleep* 16:372–86.

King, J. E. 1983. *Seals of the world*. Ithaca, NY: Cornell University Press.

Kita, H. 1965. Review of activities: Harvest, seasons, and diving patterns. In *Physiology of breath-hold diving in the Ama of Japan*, ed. H. Rahn, 41–56. Publication 1341. Washington, DC: National Academy of Sciences–National Research Council.

Kitakaze, M., M. L. Weisfeldt, and E. Marban. 1988. Acidosis during early reperfusion prevents myocardial stunning in perfused ferret hearts. *Journal of Clinical Investigation* 82:920–27.

Kitakaze, M., S. Takashima, H. Funaya, T. Minamino, K. Node, Y. Shinozaki, H. Mori, and M. Hori. 1997. Temporary acidosis during reperfusion limits myocardial infarct size in dogs. *American Journal of Physiology* 272:H2071–78.

Kjekshus, J. K., A. S. Blix, R. Elsner, R. Hol, and E. Amundsen. 1982. Myocardial blood flow and metabolism in the diving seal. *American Journal of Physiology* 242:R79–104.

Kloner, R. A. 2009. Clinical application of remote ischemic preconditioning. *Circulation* 119:776–78.

Kloner, R. A., and R. B. Jennings. 2001. Consequences of brief ischemia: Stunning, preconditioning and their clinical implications, *Circulation*, part 1, 104:2981–89; part 2, 3158–67.

Kodama, A. M., R. Elsner, and N. Pace. 1977. Effects of growth, diving history, and high altitude on blood oxygen capacity in harbor seals. *Journal of Applied Physiology* 42:852–58.

Koestler, A. 1960. *The lotus and the robot*. New York: Harper and Row.

Kooyman, G. L. 1968. An analysis of some behavorial and physiological characteristics related to diving in the Weddell seal. In *Biology of the Antarctic Seas III*, ed. W. L. Schmitt and G. A. Llano, 227–61. Washington, DC: American Geophysical Union.

————. 1989. *Diverse divers*. Berlin: Springer-Verlag.

Kooyman, G. L., and W. B. Campbell. 1973. Heart rates in freely diving Weddell seals, *Leptonychotes weddelli*. *Comparative Biochemistry and Physiology* 43:31–36.

Kooyman, G. L., D. H. Kerem, W. B. Campbell, and J. J. Wright. 1973. Pulmonary gas exchange in freely diving Weddell seals, *Leptonychotes weddelli*. *Respiration Physiology* 17:283–90.

Kooyman, G. L., E. A. Wahrenbrock, M. A. Castellini, R. W. Davis, and E. E. Sinnett. 1980. Aerobic and anaerobic metabolism during voluntary diving in Weddell seals: Evidence of preferred pathways from blood chemistry and behavior. *Journal of Comparative Physiology* 138:335–46.

Kornberger, E., and P. Mair. 1996. Important aspects in the treatment of severe accidental hypothermia: The Innsbruck experience. *Journal of Neurosurgical Anesthesiology* 8:83–87.

Kothari, L. K., A. Bordia, and O. P. Gupta. 1973. Studies on a yogi during an eight-day confinement in a sealed underground pit. *Indian Journal of Medical Research* 61:1645–50.

Kreider, M. B., E. R. Buskirk, and D. E. Bass. 1958. Oxygen consumption and body temperatures during the night. *Journal of Applied Physiology* 12:361–66.

Kreider, M. B., and P. F. Iampietro. 1959. Oxygen consumption and body temperature during sleep in cold environments. *Journal of Applied Physiology* 14:765–67.

Krogh, A. 1941. *The anatomy and physiology of respiratory mechanisms*. Philadelphia: University of Pennsylvania Press.

Kroncke, G. M., R. D. Nicols, J. T. Mendenhall, P. D. Myerowitz, and J. R. Starling. 1986. Ectothermic philosophy of acid-base balance to prevent fibrillation during hypoxia. *Archives of Surgery* 121:303–4.

Kuzuya, T., S. Hoshida, N. Yamashita, H. Fuji, M. Oe, M. Hori, T. Kamada, and M. Tada. 1993. Delayed effects of sublethal ischemia on the acquisition of tolerance to ischemia. *Circulation Research* 72:1293–99.

Larkin, J. E., and H. C. Heller. 1999. Sleep after arousal from hibernation is not homeostatically regulated. *American Journal of Physiology* 276:R522–29.

Lavigne, D. M., S. Innes, G. A. Worthy, K. M. Kovacs, O. J. Schmitz, and J. P. Hickie. 1986. Metabolic rates of seals and whales. *Canadian Journal of Zoology* 64:279–84.

Le Boeuf, B. J., D. P. Costa, A. C. Huntley, and S. D. Feldkamp. 1988. Continuous deep diving in female northern elephant seals, *Mirounga angustirostris*. *Canadian Journal of Zoology* 66:446–58. f

Lenfant, C., R. Elsner, G. L. Kooyman, and C. M. Drabek. 1969. Respiratory function of blood of the adult and fetal Weddell seal *Leptonychotes weddelli*. *American Journal of Physiology* 216:1595–97.

Liem, D. A., P. D. Verdouw, H. Ploeg, S. Kazim, and D. J. Duncker. 2002. Sites of action of adenosine in interorgan preconditioning of the heart. *American Journal of Physiology* 283:H29–37.

Liggins, G. C., J. Qvist, P. W. Hochachka, B. J. Murphy, R. K. Creasy, R. C. Schneider, M. T. Snider, and W. M. Zapol. 1980. Fetal cardiovascular and metabolic responses to simulated diving in the Weddell seal. *Journal of Applied Physiology* 49:424–30.

Lin, Y.-C. 1990. Physiological limitations of humans as breath-hold divers. In *Man in the Sea*, ed. Y.-C. Lin and K. S. Shida, 2:33–56. San Pedro, CA: Best.

Ludman, A. J., D. M. Yellon, and D. J. Hausenloy. 2010. Cardiac preconditioning for ischemia: Lost in translation. *Disease Models and Mechanisms* 3:35–38.

Lutz, A., H. A. Slagter, J. D. Dunne, and R. J. Davidson. 2008. Attention regulation and monitoring in meditation. *Trends in Cognitive Sciences* 12:163–69.

Lutz, P. L., G. E. Nilsson, and M. A. Peréz-Pinzón. 1996. Anoxia tolerant animals from a neurological perspective. *Comparative Biochemistry and Physiology B* 113:3–13.

Lutz, P. L., G. E. Nilsson, and H. M. Prentice. 2003. *The brain without oxygen: Causes of failure; physiological and molecular mechanisms for survival.* 3rd ed. Dordrecht: Kluwer Academic Publishers.

Malan, A. 1988. pH and hypometabolism in mammalian hibernation. *Canadian Journal of Zoology* 66:95–98.

———. 1999. Acid-base regulation in hibernation and aestivation. In *Regulation of acid-base status in animals and plants: An appraisal of current techniques,* ed. S. Egginton, E. W. Taylor, and J. A. Raven, 323–40. Cambridge: Cambridge University Press.

Margulis, L,. and D. Sagan. (1986) 1997. *Microcosmos: Four billion years of microbial evolution.* Berkeley and Los Angeles: University of California Press.

Masuda, Y., A. Yoshida, F. Hayashi, K. Sasaki, and Y. Honda. 1982. Attenuated ventilatory responses to hypercapnia and hypoxia in assisted breath-hold divers (funado). *Japanese Journal of Physiology* 32:327–36.

McCord, J. M. 1985. Oxygen-derived free radicals in postischemic tissue injury. *New England Journal of Medicine* 312:354–64.

McGrath, J. J., and R. W. Bullard. 1968. Altered myocardial performance in response to anoxia after high-altitude exposure. *Journal of Applied Physiology* 25:761–64.

Meerson, F. Z., O. A. Gomzakov, and M. V. Shimkovich. 1973. Adaptation to high altitude hypoxia as a factor preventing development of myocardial ischemic necrosis. *American Journal of Cardiology* 31:30–34.

Meir, J. U., and P. J. Ponganis. 2010. Blood temperature profiles of diving seals. *Physiological and Biochemical Zoology* 83:531–50.

Meir, J. U., C. D. Champagne, D. P. Costa, C. L. Williams, and P. J. Ponganis. 2009. Extreme hypoxemic tolerance and blood oxygen depletion in diving elephant seals. *American Journal of Physiology* 297:R927–39.

Meiselman, H. J., M. A. Castellini, and R. Elsner. 1992. Hemorheological behavior of seal blood. *Clinical Hemorheology* 12:657–75.

Metcalfe, J., W. Moll, and H. Bartels. 1964. Gas exchange across the placenta. *Federation Proceedings* 23:774–80.

Milano, G., P. Bianciardi, A. F. Corno, E. Raddatz, S. von Morel, L. K. von Segesser, and M. Samaja. 2004. Myocardial impairment in chronic hypoxia is abolished by short aeration episodes: Involvement of K+ATP channels. *Experimental Biology and Medicine* 229:1196–1205.

Milano, G., A. F. Corno, S. Lippa, L. K. von Segesser, and M. Samaja. 2002. Chronic and intermittent hypoxia induce different degrees of myocardial tolerance to hypoxia-induced dysfunction. *Experimental Biology and Medicine* 227:389–97.

Miller, N. E. 1969. Learning of visceral and glandular responses. *Science* 163:434–45.

Mitz, S. A., S. Reuss, L. P. Folkow, A. S. Blix, J. M. Ramirez, T. Hankeln, and T. Burmester. 2009. When the brain goes diving: Glial oxidative metabolism may confer tolerance to the seal brain. *Neuroscience* 163:552–60.

Mohri, M., R. Torii, K. Nagaya, K. Shiraki, R. Elsner, H. Takeuchi, Y. S. Park, and S. K. Hong. 1995. Diving patterns of ama divers of Hegura Island, Japan. *Undersea Medical Research* 22:137–43.

Monaghan, P., and E. G. Viereck. 1999. *Meditation: The complete guide.* Novato, CA: New World Library.

Mortola, J. P. 1993. Hypoxic hypometabolism in mammals. *News in Physiological Sciences* 8:79–82.

———. 1999. How newborn mammals cope with hypoxia. *Respiration Physiology* 116: 95–103.

———. 2004. Implications of hypoxic hypometabolism during mammalian ontogenesis. *Respiration Physiology and Neurobiology* 141:345–56.

Mortola, J. P., R. Rezzonico, and C. Lanthier. 1989. Ventilation and oxygen consumption during acute hypoxia in newborn mammals: A comparative analysis. *Respiration Physiology* 78:31–43.

Muir, T. J., J. P. Costango, and R. E. Lee Jr. 2008. Metabolic depression induced by urea in organs of the wood frog, *Rana sylvatica*: Effects of season and temperature. *Journal of Experimental Zoology A: Ecological Genetics and Physiology* 309:111–16.

Murry, C. E., R. B. Jennings, and K. A. Reimer. 1986. Preconditioning with ischemia: A delay of lethal cell injury in ischemic myocardium. *Circulation* 74:1124–36.

Murry, C. E., V. J. Richard, K. A. Reimer, and R. B. Jennings. 1990. Ischemic preconditioning slows energy metabolism and delays ultrastructural damage during a sustained ischemic episode. *Circulation Research* 66:913–31.

Naranjo, C., and R. E. Ornstein. 1971. *On the psychology of meditation.* New York: Viking Press.

Neubauer, J. A. 2001. Physiological and pathophysiological responses to intermittent hypoxia. *Journal of Applied Physiology* 90:1593–99.

Nilsson, G. E., and P. L. Lutz. 2004. Anoxia tolerant brains. *Journal of Cerebral Blood Flow and Metabolism* 24:475–86.

Noren, S. R., and T. M. Williams. 2000. Body size and skeletal muscle myoglobin of cetaceans: Adaptations for maximizing dive duration. *Comparative Biochemistry and Physiology A* 126:181–91.

Nystul, T. G., and M. B. Roth. 2004. Carbon monoxide-induced suspended animation protects against hypoxic damage in *Caenorhabditis elegans*. *Proceedings of the National Academy of Sciences of the USA* 101:9133–36.

Ostadal, B., and F. Kolar. 2007. Cardiac adaptation to chronic high-altitude hypoxia: Beneficial and adverse effects. *Respiration Physiology and Neurobiology* 158:224–36.

Ostadal, B., I. Ostadalova, and N. S. Dhalla. 1999. Development of cardiac sensitivity to oxygen deficiency: Comparative and ontogenetic aspects. *Physiological Reviews* 79: 635–59.

Park, Y. S., K. Shiraki, and S. K. Hong. 1983. Energetics of breath-hold diving in Korean and Japanese professional divers. *Undersea Biomedical Research* 10:203–15.

Patel, B., R. A. Kloner, K. Przyklenk, and E. Braunwald. 1988. Postischemic myocardial "stunning": A clinically relevant phenomenon. *Annals of Internal Medicine* 108:626–29.

Paulev, P. 1965. Decompression sickness following repeated breath-hold dives. *Journal of Applied Physiology* 20:1028–31.

Pell, T. J., G. F. Baxter, D. M. Yellon, and G. M. Drew. 1998. Renal ischemia preconditions myocardium: Role of adenosine receptors and ATP-sensitive potassium channels. *American Journal of Physiology* 275:H1542–47.

Polasek, L. K., K. A. Dickson, and R. W. Davis. 2006. Metabolic indicators in the skeletal muscles of harbor seals (*Phoca vitulina*). *American Journal of Physiology* 290:R1720–27.

Ponganis, P. J., T. K. Stockard, D. H. Levenson, L. Berg, and E. A. Baranov. 2006. Intravascular pressure profiles in elephant seals: Hypotheses on the caval sphincter, extradural vein and venous return to the heart. *Comparative Biochemistry and Physiology A* 145:123–30.

Ponganis, P. J., G. L. Kooyman, and S. H. Ridgway. 2003. Comparative diving physiology. In *Physiology and medicine of diving*, ed. A. O. Brubakk and T. S. Neuman, 211–26. Edinburgh: Saunders.

Ponganis, P. J., U. Kreutzer, T. K. Stockard, P. C. Lin, N. Sailasuta, T. K. Tran, R. Hurd, and T. Jue. 2008. Blood flow and metabolic regulation in seal muscle during apnea. *Journal of Experimental Biology* 211:3323–32.

Ponganis, P. J., J. U. Meir, and C. L. Williams. 2011. In pursuit of Irving and Scholander: A review of oxygen store management in seals and penguins. *Journal of Experimental Biology* 214:3325–39.

Poupa, O. 1976. Some trends of the natural defense against the cardiac anoxia. *Acta Medica Scandinavica Supplementum* 587:47–56.

Poupa, O., K. Krofta, J. Prochazka, and Z. Turek. 1966. Acclimation to simulated high altitude and acute cardiac necrosis. *Federation Proceedings* 25:1243–46.

Prosser, C. L., L. M. Barr, R. D. Pinc, and C. Y. Lauer. 1957. Acclimation of goldfish to low concentrations of oxygen. *Physiological Zoology* 30:137–41.

Qvist, J., R. D. Hill, R. C. Schneider, K. J. Falke, G. C. Liggins, M. Guppy, R. L. Eliot, P. W. Hochachka, and W. M. Zapol. 1986. Hemoglobin concentrations and blood gas tensions of free-diving Weddell seals. *Journal of Applied Physiology* 61:1560–69.

Rahimtoola, S. H. 1989. The hibernating myocardium. *American Heart Journal* 117:211–21.

Rahn, H., ed. 1965. *Physiology of breath-hold diving and the Ama of Japan*. Washington, DC: National Academy of Sciences, National Research Council, pub. 1341.

Rahn, H., R. B. Reeves, and B. J. Howell. 1975. Hydrogen ion regulation, temperature, and evolution. *American Review of Respiratory Disease* 112:165–72.

Ramirez, J-M., L. P. Folkow, and A. S. Blix. 2007. Hypoxia tolerance in mammals and birds: From the wilderness to the clinic. *Annual Review of Physiology* 69:113–43.

Reed, J. Z., P. J. Butler, and M. A. Fedak. 1994. The metabolic characteristics of locomotory muscles of grey seals (*Halichoerus grypus*), harbor seal (*Phoca vitulina*), and Antarctic fur seals (*Arctocephalus gazella*). *Journal of Experimental Biology* 194:33–46.

Reimer, K. A., R. S. Vander Heide, and R. B. Jennings. 1994. Ischemic preconditioning slows ischemic metabolism and limits myocardial infarct size. *Annals of the New York Academy of Sciences* 723:99–115.

Rennie, D. W. 1965. Thermal insulation of Korean diving women and non-divers in water. In Rahn 1965, 315–24.

Rennie, D. W., B. G. Covino, B. J. Howell, S. H. Song, B. S. Kang, and S. K. Hong. 1962. Physical insulation of Korean diving women. *Journal of Applied Physiology* 17: 961–66.

Richet, C. 1899. De la résistance des canards á l'asphyxie. *Journal de Physiologie et Pathologie General* 1:641–50.

Ridgway, S. H., D. A. Carter, and W. Clark. 1975. Conditioned bradycardia in the sea lion *Zalophus californianus*. *Nature* 256:37–38.

Ridgway, S. H., B. L. Scronce, and J. Kanwisher. 1969. Respiration and deep diving in the bottlenose porpoise. *Science* 166:1651–54.

Rohlicek, C. V., C. Saiki, T. Matsuoka, and J. P. Mortola. 1998. Oxygen transport in conscious newborn dogs during hypoxic hypometabolism. *Journal of Applied Physiology* 84:763–68.

Roth, M. B., and T. Nystul. 2005. Buying time in suspended animation. *Scientific American* 292 (6): 48–55.

Rowell, L. B. 1974. Human cardiovascular adjustments to exercise and thermal stress. *Physiological Reviews* 54:75–159.

Rubia, K. 2009. The neurobiology of meditation and its clinical effectiveness in psychiatric disorders. *Biological Psychology* 82:1–11.

Ruby, N. F. 2003. Hibernation: When good clocks go cold. *Journal of Biological Rhythms* 18:275–86.

Schaefer, K. E. 1965. Adaptation to breath-hold diving. In Rahn 1965, 237–52.

Schagatay, E., J. P. A. Andersson, M. Hallén and B. Pålsson. 2001. Physiological and genomic consequences of intermittent hypoxia: Selected contribution: Role of spleen emptying in prolonging apneas in humans. *Journal of Applied Physiology* 90:1623–29.

Schagatay, E., H. Haughey, and J. Reimers. 2005. Speed of spleen volume changes evoked by serial apneas. *European Journal of Applied Physiology* 93:447–52.

Schagatay, E., M. X. Richardson, and A. Lodin-Sundström. 2012. Size matters: spleen and lung volumes predict performance in human apneic divers. *Frontiers in Physiology* 3:1–8.

Schmidt, M. R., M. Smerup, I. E. Konstantinov, M. Shimizu, J. Li, M. Cheung, P. A. White, S. B. Kristiansen, K. Sorensen, V. Dzavik, A. N. Redington, and R. K. Kharbanda. 2007. Intermittent peripheral tissue ischemia during coronary ischemia reduces myocardial infarction through a KATP-dependent mechanism: First demonstration of remote ischemic perconditioning. *American Journal of Physiology* 292:H1883–90.

Scholander, P. F. 1940. Experimental investigations on the respiratory function in diving mammals and birds. *Hvalrådets Skrifter*, Oslo, 22:1–131.

———. 1963. The master switch of life. *Scientific American* 209 (6): 92–106.

Scholander, P. F., L. Irving, and S. W. Grinnell. 1942a. Aerobic and anaerobic changes in seal muscles during diving. *Journal of Biological Chemistry* 142:431–40.

———. 1942b. On the temperature and metabolism of the seal during diving. *Journal of Cellular and Comparative Physiology* 19:67–78.

Scholander, P. F., H. T. Hammel, K. Lange Andersen, and Y. Løyning. 1958. Metabolic acclimation to cold in man. *Journal of Applied Physiology* 12:1–8.

Sekar, T. S., K. F. MacDonnell, P. Namsirikul, and R. S. Herman. 1980. Survival after prolonged submersion in cold water without neurologic sequelae: Report of two cases. *Archives of Internal Medicine* 140:775–79.

Semenza, G. L. 2000. HIF-1: Mediator of physiological and pathophysiological responses to hypoxia. *Journal of Applied Physiology* 88:1474–80.

————. 2007. Life with oxygen. *Science* 318:62–64.

Sequeira, S., ed. 2014. *Advances in meditation research: Neuroscience and clinical applications.* Vol. 1307 of *Annals of the New York Academy of Sciences.*

Sharbaugh, S. M. 2001. Seasonal acclimatization to extreme climatic conditions by black-capped chickadees (*Poecile atricapilla*) in Interior Alaska (64°N). *Physiological and Biochemical Zoology* 74:568–75.

Shizukuda, N., R. T. Mallet, S. Lee, and H. F. Downey. 1992. Hypoxic preconditioning of ischaemic canine myocardium. *Cardiovascular Research* 26:534–42.

Sidi, D., J. R. G. Kuipers, D. Teitel, M. A. Heymann, and A. M. Rudolph. 1983. Developmental changes in oxygenation and circulatory responses to hypoxemia in lambs. *American Journal of Physiology* 245:H674–82.

Siebeke, H., H. Breivik, T. Rød, and B. Lind. 1975. Survival after 40 minutes' submersion without cerebral sequelae. *Lancet* 305:1275–77.

Singer, D. 1999. Neonatal tolerance to hypoxia: A comparative-physiological approach. *Comparative Biochemistry and Physiology A* 123:221–34.

Smith, H. 1991. *The world's religions: Our great wisdom traditions.* New York: HarperSan-Francisco.

Snapp, B. D., and H. C. Heller. 1981. Suppression of metabolism during hibernation in ground squirrels (*Citellus lateralis*). *Physiological Zoology* 54:297–307.

Solberg, E. E., O. Ekeberg, A. Holen, F. Ingjer, L. Sandvik, P. A. Standal, and A. Vikman. 2004. Hemodynamic changes during long meditation. *Applied Psychophysiology and Biofeedback* 29:213–21.

Sordahl, L. A., G. Mueller, and R. Elsner. 1983. Comparative functional properties of mitochondria from seal and dog hearts. *Journal of Molecular and Cellular Cardiology* 15:1–5.

Sparling, C. E., and M. A. Fedak. 2004. Metabolic rates of captive grey seals during voluntary diving. *Journal of Experimental Biology* 207:1615–24.

Stănescu, D. C., B. Nemery, C. Veriter, and C. Maréchal. 1981. Pattern of breathing and ventilatory response to CO_2 in subjects practicing hatha-yoga. *Journal of Applied Physiology* 51:1625–29.

Steller, G. W. 1751. *De Bestiis Marinis.* In Latin, *Nov. Comm. Acad. Sci. Petropolitanae* 2:289–398. Trans. W. Miller and J. E. Miller in *The fur seals and fur seal islands of the North Pacific Ocean*, part 3, ed. D. S. Jordan, 179–218. US Treasury Dept. doc. no. 2017. Washington, DC: USGPO, 1899.

Storey, K. B., and J. M. Storey. 1990. Metabolic rate depression and biochemical adaptation in anaerobiosis, hibernation and estivation. *Quarterly Review of Biology* 65:145–74.

Sun, H. Y., N. P. Wang, F. Kerendi, M. Halkos, H. Kin, R. A. Guyton, J. Vinten-Johansen, and Z. Q. Zhao. 2005. Hypoxic postconditioning reduces cardiomyocyte loss by inhibiting ROS generation and intracellular Ca^{++} overload. *American Journal of Physiology* 288:H1900–1908.

Taylor, M. J., J. E. Bailes, A. M. Elrifai, S. R. Shih, E. Teeple, M. L. Leavitt, J. G. Baust, and J. C. Maroon. 1995. A new solution for life without blood: Asanguineous low-flow perfusion of a whole-body perfusate during 3 hours of cardiac arrest and profound hypothermia. *Circulation* 91:431–44.

Thompson, D., and M. A. Fedak. 1993. Cardiac responses of grey seals during diving at sea. *Journal of Experimental Biology* 174:139–64.

Thoresen, M., and J. Wyatt. 1997. Keeping a cool head, post-hypoxic hypothermia—an old idea revisited. *Acta Paediatrica* 86:1029–33.

Toffler, A. 1970. *Future Shock.* New York: Bantam.

Tøien, Ø., J. Blake, D. M. Edgar, D. A. Grahn, H. C. Heller, and B. M. Barnes. 2011. Hibernation in black bears: Independence of metabolic suppression from body temperature. *Science* 331:906–9.

Tsang, A., D. J. Hausenloy, and D. M. Yellon. 2005. Editorial: Myocardial postconditioning: Reperfusion injury revisited. *American Journal of Physiology* 289:H2–7.

Van Citters, R. L., D. L. Franklin, O. A. Smith Jr., Watson, N. W., and R. Elsner. 1965. Cardiovascular adaptations to diving in the northern elephant seal *Mirounga angustirostris. Comparative Biochemistry and Physiology* 16:267–76.

Vannucci, R. C. 1990. Experimental biology of cerebral hypoxia-ischemia: Relation to perinatal brain damage. *Pediatric Research* 27:317–26.

Vannucci, R. C., and T. E. Duffy. 1974. Influence of birth on carbohydrate and energy metabolism in rat brain. *American Journal of Physiology* 226:933–40.

Vannucci, R. C., and D. J. Mujsce. 1992. Effect of glucose on perinatal hypoxic-ischemic brain damage. *Biology of the Neonate* 62:215–24.

Vázquez-Medina, J. P., N. O. Olguín-Monroy, P. D. Maldonado, A. Santamaria, M. Königsberg, R. Elsner, M. O. Hammill, J. M. Burns, and T. Zenteno-Savín. 2011. Maturation increases superoxide radical production without increasing oxidative damage in the skeletal muscle of hooded seals (*Cystophora cristata*). *Canadian Journal of Zoology* 89:206–12.

Vázquez-Medina, J. P., J. G. Soñanez-Organis, J. M. Burns, T. Zenteno-Savín, and R. M. Ortiz. 2011. Antioxidant capacity develops with maturation in the deep-diving hooded seal. *Journal of Experimental Biology* 214:2903–10.

Vázquez-Medina, J. P., T. Zenteno-Savín, and R. Elsner. 2006. Antioxidant enzymes in ringed seal tissues: potential protection against dive-associated ischemia/reperfusion. *Comparative Biochemistry and Physiology C* 142:198–204.

———. 2007. Glutathione protection against dive-associated ischemia-reperfusion in ringed seals. *Journal of Experimental Marine Biology and Ecology* 345:110–18.

Vázquez-Medina, J. P., T. Zenteno-Savín, R. Elsner, and R. M. Ortiz. 2012. Coping with physiological oxidative stress: a review of antioxidant strategies in seals. *Journal of Comparative Physiology B* 182:741–50.

Villafuerte, F. C., R. Cardenas, and C. Monge-C. 2004. Optimal hemoglobin concentration and high altitude: A theoretical approach for Andean men at rest. *Journal of Applied Physiology* 96:1581–88.

Vinten-Johansen, J., Z. Q. Zhao, R. Jiang, A. J. Zatta, and G. P. Dobson. 2007. Preconditioning and postconditioning: Innate cardioprotection from ischemia-reperfusion injury. *Journal of Applied Physiology* 103:1441–48.

Von Bonin, G. 1937. Brain-weight and body-weight of mammals. *Journal of General Psychology* 16:379–89.

Vreeland, R. H., W. D. Rosenzweig, and D. W. Powers. 2000. Isolation of a 250 million-year-old halotolerant bacterium from a primary salt crystal. *Nature* 407:897–900.

Walford, R. L., and S. R. Spindler. 1997. The response to calorie restriction in mammals shows features also common to hibernation: A cross-adaptation hypothesis. *Journal of Gerontology* 52A:B179–83.

Wallace, R. K. 1970. Physiological effects of transcendental meditation. *Science* 167: 1251–54.

Wallace, R. K., and H. Benson. 1972. The physiology of meditation. *Scientific American* 226 (2): 84–90.

Wallace, R. K., H. Benson, and A. F. Wilson. 1971. A wakeful hypometabolic physiological state. *American Journal of Physiology* 221:795–99.

Walpoth, B. H., and H. A. M. Daanen. 2006. Immersion hypothermia. In *Handbook on drowning: Prevention, rescue, treatment,* ed. J. Bierens, 481–531. Berlin: Springer.

Wang, L. C. H., and M. W. Wolowyk. 1988. Torpor in mammals and birds. *Canadian Journal of Zoology* 66:133–37.

Wasser, J. S. 1996. Maintenance of cardiac function during anoxia in turtles: From cell to organism. *Comparative Biochemistry and Physiology B* 113:15–22.

Watson, R. R., S. B. Kanatous, D. F. Cowan, J. W. Wen, V. C. Han, and R. W. Davis. 2007. Volume density and distribution of mitochondria in harbor seal (*Phoca vitulina*) skeletal muscle. *Journal of Comparative Physiology B* 177:89–98.

Wenger, R. H. 2000. Mammalian oxygen sensing, signalling and gene regulation. *Journal of Experimental Biology* 203:1253–63.

West, J. B., P. H. Hackett, K. H. Maret, J. S. Milledge, R. M. Peters Jr., C. J. Pizzo, and R. M. Winslow. 1983. Pulmonary gas exchange on the summit of Mount Everest. *Journal of Applied Physiology* 55:678–87.

White, F. N. 1981. A comparative physiological approach to hypothermia. *Journal of Thoracic and Cardiovascular Surgery* 82:821–31.

White, D. P., J. V. Weil, and C. W. Zwillich. 1985. Metabolic rate and breathing during sleep. *Journal of Applied Physiology* 59:384–91.

White, F. C., R. Elsner, D. Willford, E. Hill, and E. Merhoff. 1990. Responses of harbor seal and pig heart to progressive and acute hypoxia. *American Journal of Physiology* 259: R849–56.

Wickham, L. L., R. M. Bauersachs, R. B. Wenby, S. Sowemimo-Coker, H. J. Meiselman, and R. Elsner. 1990. Red cell aggregation and viscoelasticity of blood from seals, swine and man. *Biorheology* 27:191–204.

Wilhelm Filho, D., F. Sell, L. Ribeiro, M. Ghislandi, F. Carrasquedo, C. G. Fraga, J. P. Wallauer, P. C. Simões-Lopes, and M. M. Uhart. 2002. Comparison between the antioxidant status of terrestrial and diving mammals. *Comparative Biochemistry and Physiology A* 133:885–92.

Williams, T. M. 2001. Intermittent swimming by mammals: A strategy for increasing energetic efficiency during diving. *American Zoologist* 41:166–76.

Williams, T. M., J. E. Haun, and W. A. Friedl. 1999. The diving physiology of bottlenose dolphins (*Tursiops truncatus*): I. Balancing the demands of exercise for energy conservation at depth. *Journal of Experimental Biology* 202:2739–48.

Williams, T. M., and G. A. J. Worthy. 2002. Anatomy and physiology: The challenge of aquatic performance. In *Marine mammal biology: An evolutionary approach*, ed. A. R. Hoelzel, 73–97. Malden, MA: Blackwell Science.

Williams, T. M., M. Zavanelli, M. A. Miller, R. A. Goldbeck, M. Morledge, D. Casper, D. A. Pabst, W. McLellan, L. P. Cantin, and D. S. Kliger. 2008. Running, swimming and diving modifies neuroprotecting globins in the mammalian brain. *Proceedings of the Royal Society B: Biological Sciences* 275:751–58.

Williams, T. M., R. W. Davis, L. A. Fuiman, J. Francis, B. J. Le Boeuf, M. Horning, J. Calambokidis, and D. A. Croll. 2000. Sink or swim: Strategies for cost-efficient diving by marine mammals. *Science* 288:133–37.

Wood, S. C. 1991. Interactions between hypoxia and hypothermia. *Annual Review of Physiology* 53:71–85.

Wood, S. C., and R. Gonzales. 1996. Hypothermia in hypoxic animals: mechanisms, mediators, and functional significance. *Comparative Biochemistry and Physiology B* 113:37–43.

Wu, X., T. Drabek, P. M. Kochanek, J. Henchir, S. W. Stezoski, K. Cochran, R. Garman, and S. A. Tisherman. 2006. Induction of profound hypothermia for emergency preservation and resuscitation allows intact survival after cardiac arrest resulting from prolonged lethal hemorrhage and trauma in dogs. *Circulation* 113:1974–82.

Wyatt, J. S., and M. Thoresen. 2011. Hypothermia treatment and the newborn. *Pediatrics* 100:1028–29.

Xie, Y., W. Z. Zhu, Y. Zhu, L. Chen, Z. N. Zhou, and H. T. Yang. 2004. Intermittent high altitude hypoxia protects the heart against lethal Ca^{2+} overload injury. *Life Sciences* 76:559–72.

Yellon, D. M., and J. M. Downey. 2003. Preconditioning the myocardium: From cellular physiology to clinical cardiology. *Physiological Reviews* 83:1113–51.

Young, J. D., and E. Taylor. 1998. Meditation as a voluntary hypometabolic state of biological estivation. *News in Physiological Sciences* 13:149–53.

Zachariassen, K. E. 1985. Physiology of cold tolerance in insects. *Physiological Reviews* 65:799–832.

Zapol, W. M. 1987. Diving adaptations of the Weddell seal. *Scientific American* 256 (6):100–105.

———. 1996. Diving physiology of the Weddell seal. In *Handbook of physiology*, sec. 4, *Environmental physiology*, ed. M. J. Fregly and C. M. Blatteis, 2:1049–56. New York and Oxford: American Physiological Society.

Zapol, W. M., G. C. Liggins, R. C. Schneider, J. Qvist, M. T. Snider, R. K. Creasy, and P. W. Hochachka. 1979. Regional blood flow during simulated diving in the conscious Weddell seal. *Journal of Applied Physiology* 47:968–73.

Zenteno-Savín, T., E. Clayton-Hernández, and R. Elsner. 2002. Diving seals: Are they a model for coping with oxidative stress? *Comparative Biochemistry and Physiology C* 133:527–36.

Zenteno-Savín, T., J. P. Vázquez-Medina, N. Cantú-Medellín, P. J. Ponganis, and R. Elsner. 2012. Ischemia/reperfusion in diving birds and mammals: How they avoid oxidative damage. In *Oxidative stress in aquatic ecosystems*, ed. D. Abele, J. P. Vázquez-Medina, and T. Zenteno-Savín, 178–89. Hoboken, NJ: Wiley-Blackwell.

Zhao, Z. Q., J. S. Corvera, M. E. Halkos, F. Kerendi, N. P. Wang, R. A. Guyton, and J. Vinten-Johansen. 2003. Inhibition of myocardial injury by ischemic postconditioning during reperfusion: Comparison with preconditioning. *American Journal of Physiology* 285:H579–88.

Zong, P., S. Setty, W. Sun, R. Martinez, J. D. Tune, I. V. Ehrenberg, E. N. Tkatchouk, R. T. Mallet, and H. F. Downey. 2004. Intermittent hypoxic training protects canine myocardium from infarction. *Experimental Biology and Medicine* 229:806–12.

INDEX